# THE DEMONSTRATION
# OF
# DIRECT POTABLE WATER REUSE

## THE DENVER PROJECT TECHNICAL REPORT
## (1979-1993)

### WILLIAM C. LAUER

**The Demonstration of Direct Potable Water Reuse:**
**The Denver Project Technical Report (1979-1993)**

William C. Lauer

---

**Disclaimer**

Although this report has been extensively reviewed for accuracy, the events and results represented herein occurred several decades ago.  Consequently, the author necessarily relied on printed reports and other documents that presented the information adapted for this publication.  The accuracy of these references could not be crosschecked with other information in many cases since corroborating documents no longer exist. Fortunately, the author was also the author of the previously printed original documents in most cases so firsthand knowledge could be used to augment the historic records.

This report contains excerpts from, and is largely based on, the Denver Direct Potable Water Reuse Demonstration Project Final Report by William C. Lauer for USEPA Cooperative Agreement No. CS-806821-01-4 (April 1993). Although the research described here was wholly or in part funded by the United States Environmental Protection Agency and Denver Water (Denver Board of Water Commissioners), it has not been subject to a formal review by either agency therefore it does not necessarily reflect their views and official endorsement should not be inferred.

This report is designed only to provide information on the Denver Direct Potable Water Reuse Demonstration Project. This information is provided and sold with the knowledge that the publisher and author do not offer any legal or other professional advice. In the case of a need for any such expertise consult with the appropriate professional. This report does not contain all information available on the subject. This report has not been created to be specific to any individual's or organizations' situation or needs. Every effort has been made to make this report as accurate as possible. There may still be typographical and or content errors. Thus, this report should serve only as a general guide and not as the ultimate source of subject information. This report contains information that might be dated and is intended only to educate and inform. The author and publisher shall have no liability or responsibility to any person or entity regarding any loss or damage incurred, or alleged to have incurred, directly or indirectly, by the information contained in this report.

ISBN-13: 978-1522855446
ISBN-10: 1522855440

Published in the United States of America


William C. Lauer
QualQuest, LLC

Inquiries should be sent to blauerq2@gmail.com.

# Foreword By Denver Water

Periodically, an organization takes on something so innovative – so bold – that they are decades ahead of their time. Such was the case with Denver Water's Potable Reuse Demonstration Project. Using sound science, peer-review, and collaboration with universities, public health agencies, consulting engineers and equipment manufacturers, Denver Water employees proved the feasibility of making safe drinking water from secondary-treated wastewater. The obstacles weren't small; the cost not insignificant. The team who carried out this task faced numerous uncertainties while working in areas of science that had not been fully developed.

Years later, the water industry can reflect on this project as a key milestone in the history of water treatment. It advanced methodologies that paved the way for future potable reuse projects all over the world. Experiments utilizing ultrafiltration, reverse osmosis, granular activated carbon, ozone, and ultraviolet light disinfection laid the foundation for the multi-barrier treatment approach for potable reuse considered standard today. Experiments in nutrient recovery were conducted years before the concept of recovering these resources from used water became popular. Research about public acceptance and communication techniques provided insights that still are used to craft communication strategies around potable reuse. And whole animal toxicity studies, though perhaps controversial by today's standards, went farther than any other research to prove the safety of potable reuse.

But this wasn't just a big science project. Climate change, population growth, and the demands of water by society have stressed water supplies in many parts of the world, to the point where our way of life in these areas is being impacted negatively. Traditional "clean" sources of water are no longer sufficient to meet our needs. In addition to conservation and efficiency measures, all water sources – from stormwater to seawater to wastewater – need to be evaluated and considered by communities looking to match supply with demand. We have but one water supply and must exercise wise stewardship of it to ensure there is enough for the environment, recreation, and humans. THAT is what makes Denver Water's Potable Reuse Demonstration Project so important. The project team proved that through proper application of technology and social science, used water can be safely and intentionally reused for human consumption again and again.

Our thanks go to Bill Lauer, who was the project manager for the Potable Reuse Demonstration Project and whose persistence to share the results of this research more broadly is the reason you are reading these words. Bill and the other Denver Water employees and researchers who dedicated many years of their lives to the project, embody the vision, passion and excellence Denver Water prides itself in. Thank you all.

Brian Good
Deputy Manager of Organizational Improvement, Denver Water
Past President, WateReuse Association
January 25, 2016

# Author's Comments

This report summarizes of the results from Denver's Direct Potable Water Reuse Demonstration Project. The Project, conducted between 1979 and 1993, demonstrated that potable water could be reliably produced from unchlorinated secondary treated wastewater. This was a landmark for drinking water treatment and greatly furthered the understanding of water quality. The results of this research have been used by those considering this alternative to meet drinking water needs in water-short areas around the world.

The Project scope was unprecedented and has never been duplicated. No other municipal organization has attempted to design, construct, operate, and evaluate such a complex treatment facility. The Project was conducted nevertheless with the same number of personnel with the same education and training as those that operate conventional water treatment facilities. Although consultants and many expert advisors were involved, the day-to-day operation and the engineering and scientific evaluations were conducted by Denver Water Department staff.

The results presented here are largely from the twenty-five volume Final Report approved by the United States Environmental Protection Agency (US EPA) in April 1993. That report completed the requirements under the Cooperative Agreement between the Denver Water Department and US EPA to conduct a research project to demonstrate the feasibility of direct potable water reuse. Additional detail about the Project is available by consulting the technical literature and the Project Final Report available from Denver Water and US EPA (Cooperative Agreement CS-806821-01-4).

~ ~

I would like to acknowledge the contribution of all those who were involved in this incredible project. The successful outcome could only have been achieved with everyone's commitment.

I thank all of those who provided information about the Project for presentation in this report. A special thank you is offered to my good friend and former project administrative assistant (name withheld). She worked with me for many hours helping to research the documents we saved about the Project. Thank you also to Holly Geist, Denver Water Archives Specialist, who helped us find the stored documents at Denver Water. A special thank you goes to Brian Good, Deputy Manager of Organizational Improvement for Denver Water, who provided support for the publication of this report.

And finally, I thank Denver Water. This is a great water company that gave me the opportunity of a lifetime.


William C. Lauer

# Contents

# List of Tables

# List of Tables
## (continued)

# List of Figures

# 1

# Why Denver Investigated Potable Reuse?

The Direct Potable Water Reuse Demonstration Project was conducted to examine the feasibility of reusing highly treated wastewater as drinking water to meet Denver's future water supply needs. To accomplish this objective the Project examined the issues of water quality, treatment plant reliability, potential chronic health effects, technical process operability, public acceptance, and cost. A 1 million gallon per day (mgd) demonstration plant was designed and built that housed the multitude of treatment process systems that produced test samples for the health effects and water quality studies. The facility served as the focal point for public information efforts that were aimed at increasing acceptance for direct potable reuse. The ten-year $34 million project received funding from the United States Environmental Protection Agency (US EPA contributed about 20%) and Denver Water (The Denver Board of Water Commissioners). Information gained from the Project has provided a basis for the consideration of this alternative as a possible future drinking water source for Denver.

~ ~

In 1964 the water supply for the Denver metropolitan area was adequate to meet the demands of the growing area. Projections though showed that additional supplies would be needed in the future. The South Platte River was the primary natural source of water for the metro area. This source was already insufficient to supply all the demands, and these flows had to be augmented with trans-mountain diversions.

The first major system developed to divert water to Denver from the west slope of the continental divide was the Moffat system completed in 1937. This was followed in 1964 by the Blue River system. The water supply situation for the Denver service area in the 1970's (similar today) was about 40% from the native East slope supply and 60% from trans-mountain water diverted into the Denver area.

Colorado water law generally prohibits the reuse of water. Once used, water must be returned to the nearest water-course for use by the next priority water right. But, under Colorado water law, water diverted from the west slope of the continental divide to the east slope did not need to be returned. As a result, this water can be legally reused for beneficial uses such as drinking water. As a responsible steward of trans-mountain water, Denver sought ways to reuse this valuable resource to augment its overall water supply.

Denver's Successive Use Program was initiated in 1964 in response to the Blue River decree (allowing diversion of water from the Blue River on the west slope of the continental divide). This court directive carried with it the stipulation that water derived from the west slope, in particular the water from the Blue River system, would be reused successively to meet various water demands with the understanding that this reuse would lessen the need for more diversions. The Denver's Successive Use program was designed, hence, to evaluate options to satisfy this decree.

The Successive Use Project investigated many possibilities other than potable reuse. Alternatives such as river exchange, groundwater recharge, industrial reuse, agricultural reuse, and dual distribution of non-potable water were studied besides potable reuse. Trans-mountain water used for agricultural purposes within the Denver area was prohibited by the Blue River Decree since this would likely increase diversions. Groundwater recharge was not feasible because of geologic restrictions and scattered uses. Industrial reuse and individual site lawn-irrigation were found to be uneconomical because of the low and scattered demand. Due to various legal, economic, and technical considerations, these options were then discarded in favor of potable reuse.

Two other options which still hold promise for at least part of the available water are exchange and dual distribution of non-potable reuse water. Exchanges were used to the greatest extent possible, but changes in supply and demand may open new opportunities to expand exchange and these possibilities continue being evaluated. Dual distribution of non-potable water for domestic use at individual sites generally required a new area of development of at least 10,000 population to be economically feasible.[1]

---

[1] *Non-potable reuse through a separate distribution pipe network is currently practiced by Denver Water. A 30 mgd non-potable reuse treatment plant began production in 2004. Water produced by the plant is distributed to industrial and governmental customers. This plant was not part of the water supply analysis conducted in the late 1960's or before beginning the*

Successive Use Program studies found that the non-potable uses (like landscape irrigation) that required lower treatment costs and were readily accepted by the public would not be able to use all the water that was available. Direct potable reuse was the only successive use alternative that could use the entire 100 mgd that could be reused. This finding lead to initiation of research, that began in 1968, in collaboration with the University of Colorado's Environmental Engineering Department to examine treatment processes that would be necessary to convert wastewater to drinking water.

To determine the processes and optimum operating conditions for potable reuse treatment, a pilot plant was constructed in 1970. The plant was located at the Denver Metropolitan Sewage Disposal District plant site (later renamed Denver Metro Wastewater Reclamation District) and was initially funded by a construction grant from the Federal Water Quality Administration (which was the predecessor to the US EPA). The Denver Water Department thereafter operated and financed it in collaboration with the University of Colorado Environmental Engineering Department. Over the ten years of operation more than thirty graduate degrees resulted from the studies conducted at the pilot facility.

Many treatment processes were investigated at the pilot plant to convert secondary treated wastewater to potable water quality. The results from these studies lead to the conclusion that this conversion was feasible; and although expensive, the costs were within the range projected for acquiring new trans-mountain supplies.

Plans were initiated to construct a demonstration treatment facility to continue to study the reliability and cost of potable reuse. This step between pilot studies and the construction of a full-scale treatment plant was considered necessary due to the pioneer nature of this conversion and to answer the many technical and non-technical issues which would need to be addressed before full-scale implementation. The multi-million dollar Direct Potable Water Reuse Demonstration Project was proposed to provide these answers.

---

*Demonstration Project. It is expected that the demand for non-potable reuse water will increase in the future. Alternative reuse options, including potable, are still being considered to meet Denver's projected water demand.*

# 2

# Project Plans

Denver's Direct Potable Water Reuse Demonstration Project was designed to determine the feasibility of reusing wastewater as drinking water. To achieve this goal the Project sought to establish product water safety, demonstrate treatment plant reliability, increase public awareness, improve regulatory agency awareness, and provide data for full-scale implementation. Establishing product water safety was the primary goal. Unless the unquestioned safety of the reuse product water could be established the other Project goals could not be met.

## Establish Product Water Safety

This key objective was very difficult to achieve. Since the health standards established for drinking water were not intended to apply to treated waters which were derived from polluted sources, other criteria needed to be used to ensure that the product was suitable for human consumption. Consequently, the Project included the following water quality criteria:

- The product water was compared with parameters included in the US EPA National Primary and Secondary Drinking Water Regulations and the Colorado State drinking water regulations;

- The product was compared with potential or proposed federal or state regulated parameters, World Health Organization standards and other international standards;

- The product water was compared with Denver's current drinking water in those areas where there were no existing or proposed standards; and

- Whole-animal lifetime health effects testing including chronic toxicity and carcinogenicity, and reproductive toxicity were conducted on the product water using Denver's current drinking water as a comparison standard.

5

Denver's current drinking water was selected for use as a comparison since it was derived from a relatively protected source and there was no reason to believe that it would fail to satisfy any future health standards. Meeting this criterion ensured a margin of safety necessary to apply this technology for many years hence. Also, surveys had shown that if water comparable to Denver's existing tap water supply was produced customers would be more inclined to accept it as an alternate drinking water source.

## Demonstrate the Reliability of the Process

To meet this goal the plant would need to operate continuously while producing water quality that meets all the quality criteria. This meant the Demonstration plant would operate continuously twenty-four hours a day, seven days a week simulating the operation of the Department's conventional drinking water treatment plants. Extensive water quality monitoring and animal health effects studies performed while operating the Demonstration plant over several years would further confirm the plant's reliability.

## Improve Public Awareness

Several surveys were conducted before the completion of the Demonstration plant. The last comprehensive one was completed in 1982 with the follow-up in 1985. All these opinion surveys revealed similar attitudes toward potable reuse: most customers supported the concept if the safety was assured. The same surveys revealed a tremendous information gap and lack of awareness which appeared to be responsible for some of the negative comments about potable reuse. To tackle this issue, the Demonstration Project included a public information program centered on the reuse treatment plant. A quantitative measure for satisfying this objective was established at 50,000 Denver residents being informed about potable water reuse.

## Increase Regulatory Agency Awareness

If the first two criteria were achieved, the federal and state agencies would be called upon to accept this treatment technology when full-scale implementation was initiated. Awareness of the Project and its results by potential regulators was critical to gain their future support for direct potable reuse. To satisfy this goal, the Project included representatives from federal, state, and local agencies on technical advisory committees, thus ensuring they were fully informed and

6

had input about the conduct of the Project. Additionally, the results of the scientific and engineering studies were presented in peer review technical publications and at professional conferences. This provided information for peer review and analysis by uninvolved leaders in the industry that could then add credibility to the results and enhance regulatory agency confidence.

## Provide Data for Full-Scale Implementation

The operational data gathered during the testing period would be used to satisfy this objective. Cost estimates along with operational information for full-scale implementation would provide the necessary comparisons that would be needed by planners considering this water resource alternative.

# Demonstration Plant Conceptual Design

Conceptual design of the Direct Potable Water Reuse Demonstration Plant, which was to serve as the main testing facility for the Project, design began in 1974. A design seminar held in Denver included national experts in advanced wastewater treatment and reuse. These industry leaders advised the Department and its engineering consultants on an appropriate treatment processes to accomplish the conversion to potable water. In 1975, the engineering consulting firm of CH2M-Hill prepared a conceptual design report based on the results of this conference. This report was used by the Department when it applied for federal funding of its proposed Reuse Demonstration Project later that year.

# USEPA Cooperative Agreement

The Denver Water Department began actively seeking federal participation in the Project shortly after publication of the conceptual design report to secure shared funding, provide technical expertise, and ensure credibility. After initial frustrations in this effort, the Department engaged the Colorado congressional delegation to gain authorization and the appropriations necessary for US EPA to participate. Several discussions between the Department and US EPA were needed before reaching a final Cooperative Agreement. This agreement, signed on June 8, 1979, provided for accomplishment of a $22 million program over an eight-year period. Initially,

US EPA's share was $7 million with the balance provided by the Denver Water Department. The total project cost ultimately increased to $34 million, but US EPA's total did not significantly change. The Project budget was divided into five parts for cost sharing purposes (Table 2-1).

**Table 2-1**
**Reuse Demonstration Project Costs**
**1979-1992**

| Project Elements | Total Cost ($) | US EPA Share ($) |
|---|---|---|
| Design | 1,319,059 | 346,958 |
| Construction | 18,501,107 | 4,150,000 |
| Operation | 8,419,591 | 600,000 |
| Scientific Studies | | |
| Water Quality | 3,085,363 | 750,000 |
| Animal Health Effects | 3,022,053 | 1,200,000 |
| **Project Cost** | **34,347,173** | **7,046,958** |

## Expert Committees and Advisors

Denver relied on the advice of a team of volunteer national experts and consultants, as well as input from regulatory agencies and other valued counselors (e.g. equipment suppliers, and system designers) during all of its reuse investigations. Early assessments performed under the Successive Use Program in the late 1960's benefited from the advice of a panel of experts. Some of these advisors continued to serve as it evolved into the Direct Potable Water Reuse Demonstration Project. Before the Cooperative Agreement was signed with the US EPA, a committee of experts participated in the development of the conceptual design of the proposed treatment plant. Later a Design Advisory Committee was formed comprised of these members along with several additional experts.

The US EPA Cooperative Agreement required a Project Advisory Committee that was assembled shortly after the 1979 signing. Other working committees were formed to assist with the development of the Analytical Studies Program and the Animal Health Effects Study. Many of the members of these committees also served on the Project Advisory Committee. Some members of these committees and their affiliations changed over the duration of the Project, sometimes more than once. Table 2-2 lists the members (a Who's Who of the nation's technical and scientific water experts in their respective fields) of the various expert advisory committees.

Consultants and other significant advisors included:

- CH2M-Hill
- Richard Arber, Arber Associates
- Dr. Joseph Borzelleca, Medical College of Virginia, Health Effects Studies Protocol Developer
- Dr. Lymon Condie, US EPA Toxicologist, Health Effects Program Analyst
- Westvaco, Fluidized Bed Carbon Regeneration Furnace
- Fluid Systems, Reverse Osmosis System Supplier
- USEPA Water Environmental Research Laboratory

## Table 2-2
## Expert Advisory Committee Members
(Some served for part of the Project)

C = Conceptual Design, D = Design, P = Project, A = Analytical, H = Health Effects

| Committee Member | Committees(s)* | Affiliation** |
|---|---|---|
| Dr. Elmer W. Akin | C, P | U.S. EPA-National Regional Office, Atlanta, Georgia |
| Dr. Fred Kopfler | P,A | U.S. EPA-Gulf of Mexico Program Office, Stennis Space Center, Mississippi |
| Dr. Franklyn N. Judson | P,H | Director, Denver Public Health, Denver, Colorado |
| Dr. Robert Neal | P,A,H | Center in Molecular Toxicology, Vanderbilt University School of Medicine, Nashville, Tennessee |
| Dr. Harold Walton | P,A | Chemistry Department, University of Colorado, Boulder, Colorado |
| Dr. Mark D. Sobsey | P,A | Professor of Environmental Microbiology, University of North Carolina, Chapel Hill, North Carolina |
| Dt. I. H. Suffet | P,A,H | Professor of Environmental Engineering and Science, University of California, Los Angeles, California |
| Dr. Joel S. Cohen | A | Professor of Mathematics and Computer Science, University of Denver, Denver, Colorado |
| Dr. Richard J. Bull | P,H | College of Pharmacy, Washington State University, Pullman, Washington |
| Dr. Alfred Dufour | P,A | U.S. EPA-Microbiology Section, Cincinnati, Ohio |
| Dr. Joseph F. Borzelleca | P, H | Department of Toxicology and Pharmacology, Medical College of Virginia, Richmond, Virginia |
| Dr. John Doull, M.D. | H | Department of Pharmacology and Toxicology, University of Kansas Medical Center, Kansas City, Kansas |
| Mr. David Argo | C,D,P | Black & Veatch Consulting Engineers, Santa Ana, California |
| Mr. Franklin D. Dryden | P,D | Environmental Engineer, Pasadena, California |
| Dr. K. Daniel Linstedt | C,D, P | Black & Veatch Consulting Engineers, Aurora, Colorado |
| Mr. Kip Cherotes | P | Legislative Assistant to the Honorable Patricia Schroeder, House of Representatives, Denver, Colorado |
| Mr. Jerry Biberstine | P | Colorado State Health Department, Denver, Colorado |
| Mr. Carl Brunner | P | Project Officer, Water Engineering, U.S. EPA-Research Laboratory, Cincinnati, Ohio |
| Dr. Lyman Condie | P,H | U.S. EPA and U. S. Army Proving Ground, Dugway, Utah |
| Dr. Joe Cotruvo | P,A | U.S. EPA-Office of Toxic Substances, Health & Environmental Review Division, Washington, D.C. |
| Dr. Raymond S. H. Yang | H | Professor Toxicology, Department of Environmental Medicine and Biomedical Sciences, Colorado State University, Fort Collins, Colorado |
| Mr. Carl Hamaan | C,D,P | CH2M Hill, Inc., Denver, Colorado |
| Mr. Jack Hoffbuhr | D | American Water Works Association, Denver, Colorado |
| Mr. John English (retired 1984) | P | Project Officer, Water Engineering, U.S. EPA-Research Laboratory, Cincinnati, Ohio |
| Dr. Rolland Grabbe | A | U.S. EPA, Denver, Co |
| Mr. John Tilstra | P, A | Chief Chemistry Section, EPA Region VII Lab, Denver, Co |
| Mr. John Puteney | P | Manager, Metropolitan Denver Sewage Disposal District, Denver, Colorado |
| Mr. Peter Sears | P | Legislative Assistant to the Honorable Patricia Schroeder, House of Representatives, Denver, Colorado |
| Dr. Tom Vernon | P,H | Colorado State Health Department, Denver, Colorado |
| Earl Feldman | P | Denver Water Department Citizens Advisory Committee representative |
| Mr. Gene Suhr | D,P | Vice President, CH2M Hill, Denver, Colorado |
| Russell Culp | C, D | President, Culp Wesner & Culp |
| Kathleen Gomez | P | Legislative Assistant to the Honorable Patricia Schroeder, House of Representatives, Denver, Colorado |

*Indicates primary service. Additional service on other committees may have been provided.
**Affiliation listed was as of the date of the writing of the Final Report or Phase completion report where service was provided

# 3

# Plant Design

The Denver Water Department hired an engineering firm to design the Reuse Demonstration Plant shortly after the signing of the US EPA Cooperative Agreement. The Department selected CH2M-Hill, because they had been the contractor for the conceptual design report in 1975 and had considerable experience designing other advanced water treatment plants. The plant design took one year starting in December 1979.

In accordance with the US EPA grant, an expert advisory committee (Table 2-2) was formed to comment on the preliminary design. Most members had participated as advisors during the pilot studies and during the preparation of the conceptual design report. New committee members were added when needed. The Design committee members continued to serve on the Project Advisory Committee with most staying on until the conclusion of the Project ten years later. CH2M-Hill proceeded with the preliminary design based upon the earlier concept and incorporated features necessitated by site-specific constraints and advisory committee input.

## Reuse Treatment Processes

The primary goal of the Reuse Demonstration Plant was the continuous and reliable production of safe potable water from treated wastewater. To achieve this objective the treatment process design applied the multiple safety barrier philosophy. This approach increases process reliability by incorporating an ensemble of unit process with redundancy in function to ensure that no single process was solely responsible for the removal of any single contaminant. There was, for instance, granular media filtration along with reverse osmosis membrane treatment to accomplish particle removal. Reverse osmosis also served as a backup or polishing step for organics removal achieved by activated carbon. Disinfection incorporated two oxidation steps instead of one.

The earliest definitive potable reuse treatment process sequence was contained in the conceptual design report prepared by CH2M Hill in 1975. During the next years, before grant award, this proposed treatment configuration was modified slightly, primarily as a result of new information from continued pilot plant testing by the Water Department and the University of Colorado. Additional adjustments were made as a result of negotiations between the Water Department and US EPA following the grant award in May 1979. More modifications were made during the early design and review process. These changes reduced the extensive cost escalations which were then predicted. Even faced with mounting costs the Water Department did not compromise its goal of producing potable water as nearly as its existing supplies as possible. Process redundancy remained in the treatment train, for example, overlooking the added cost.

The reuse treatment plant process design considered the raw water supply for the Reuse Plant would be unchlorinated secondary effluent from the Metropolitan Denver Wastewater treatment facility. The plant design treatment sequence, thus (Figure 3-1) included: high pH lime treatment, single or two-stage recarbonation, pressure filtration, selective ion exchange for ammonia removal, two stages of activated carbon adsorption, ozonation, reverse osmosis, air stripping, and chlorine dioxide disinfection. Major supporting processes included in the plant were: a fluidized bed carbon reactivation furnace, vacuum sludge filtration, and selective ion exchange ammonia removal and recovery process.

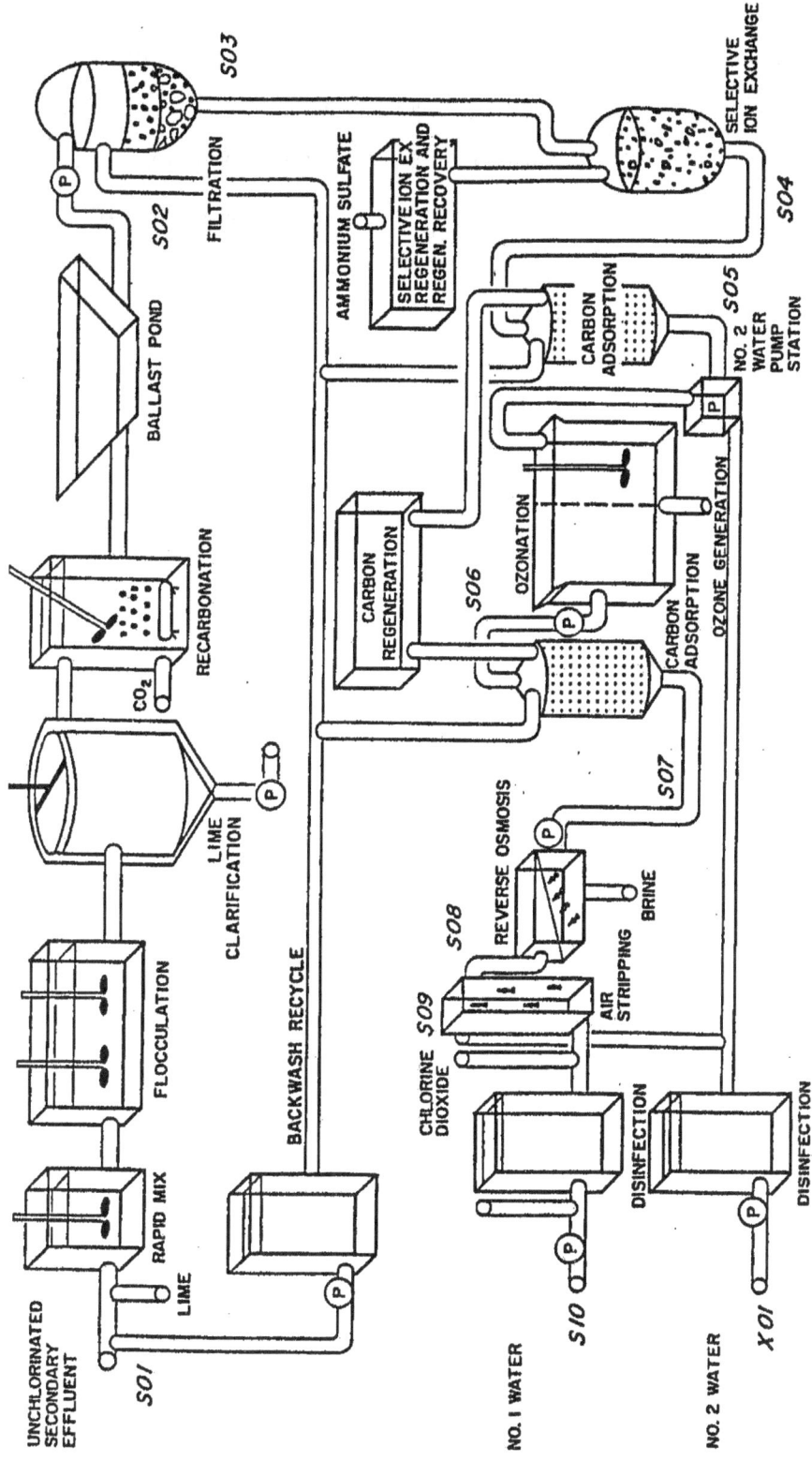

**Figure 3-1: Reuse Plant Design Treatment Sequence**

## Plant Flow Rate

The flow rate selected for the reuse treatment plant was one million gallons a day as was proposed in the conceptual design report. While this decision was reviewed from time to time this flow rate represented the lowest level that would allow realistic production sized units (for chemical treatment and filtration) to be operated in the lead processes of the treatment train. The designers recognized that smaller size chemical treatment units become toy-like and would result in process configurations different than might be used on a larger scale plant. The 1 mgd size for chemical treatment (lime clarification), filtration, ion exchange, and first-stage activated carbon, would provide operating information for realistic scale up when designing a larger facility.

The later treatment processes, following the first-stage of activated carbon treatment, were sized at 0.1 mgd. Although this was due to economics, operational data wasn't compromised because these processes were modular and so, would still provide accurate information for a full-scale future facility. The original plant design concept placed the flow reduction necessary to accomplish this downsizing after the second-stage granular activated carbon treatment (GAC). A major modification at the 10% design point moved this split to follow the first-stage activated carbon. This reduced the size and cost of the piping and contactors to be used for ozonation and second-stage GAC treatment.

## Process Redundancy

The operating philosophy, continuous operation that mimicked drinking water production, required full flow in all but unusual circumstances. A fallback position was established to operate the plant at one half flow rather than discontinuing operation completely. Provision for this 0.5 mgd capability was provided in each critical process. Thus, only the most unusual circumstances would require cessation of operation.

As a concession to cost, no backup was provided to continue operation upon a total failure of the wastewater treatment process (a very unlikely situation) or for a major power outage. Standby power was provided, nonetheless, but only enough to facilitate an orderly shut-down that would prevent damage to critical equipment.

Special provisions were incorporated into the design to prevent any inadequately treated water from progressing through the treatment process during a total power failure. These included valves that closed upon a loss of power and automatic shut-down systems where appropriate.

Because of the necessity to keep water flowing through the plant in all but the most extraordinary circumstances each unit process made use of multiple contactors. Depending upon the likelihood of an outage and the criticality of the process, this multiple contactor approach required full flow redundancy for some processes such as the lime clarification system. In other cases, two contactors of half flow capacity were each used to make up the full 1 mgd flow. In the case of filtration, for example, each of the three filters was rated at 0.5 mgd. When one filter was out of service while the second filter was being backwash the total plant flow would drop to 0.5 mgd until the backwash was completed. Most pump stations within the plant were similarly configured with three or more pumps each rated at half the maximum flow.

The two exceptions to this multiple contactor philosophy were the ozonation and disinfection processes. In each case there was only a single contact basin rated at full flow. These processes could still be interrupted for maintenance by relying on storage in the ballast ponds while they received service. There were two ballast ponds each with about 200,000 gallons capacity. This storage could support operation for about 4.8 hours at full flow. The ballast ponds also had aeration installed to maintain high levels of oxygen. This feature could be used to maintain water quality during an extended flow interruption.

The unchlorinated secondary treated wastewater from the adjacent Metropolitan Denver Sewage Disposal District was pumped to the plant using one of two 1 mgd supply pumps. Laid in the same trench as the source water pipeline were the sludge return line, and the product water return line. The Reuse Plant was designed for zero discharge so National Pollutant Discharge Elimination System (NPDES) permits were not necessary.

The Reuse Treatment Plant was to be operated as an on-line facility, like a conventional drinking water treatment plant. This required a comprehensive corrective and preventative maintenance program. The plant, thus, included a maintenance shop area and facilities for servicing the multitude of instruments and controls. A third of the plant operators performed maintenance and instrument repairs to ensure accurate information and continuous operation.

## Plant Instrumentation and Control

The treatment plant design incorporated an integrated instrumentation and control system that included both analog and digital signals needed to monitor and operate the complex treatment system. A total of 250 analog and 550 digital points were scanned by the plant process computer system every 4 seconds and the data stored in the dedicated Digital Equipment Systems® PDP 11/44 main frame mini-computer. This system monitored and controlled the more than 80 pumps, 100 motors, and 500 instruments needed for treatment plant operation. All these controls terminated in the plant control room where remote terminal units (RTU) captured the signals for processing. Most plant adjustments could be made from the control room instrument panel and monitored from the computer system monitor displays. Two display terminals were located in the plant control room and others were located in the plant visitor area, the computer room, plant laboratory and at the Water Department Quality Control Laboratory (used for data entry).

## Scientific Studies

A comprehensive water quality assessment and animal health effects study comprised the Project scientific studies. To facilitate these critical evaluations the treatment plant included several special features. Flowing taps delivered water from between each pair of unit processes into a 1200 ft.² laboratory. The in-plant laboratory combined with the Water Department's existing quality control laboratory (located about 20 miles from the Reuse Plant) provided the vast majority of the testing to support the water quality assessment program. In addition, a second 1200 ft.² area was provided in the Reuse Plant for animal health effects study sample preparation. Since the method and procedures for the animal health effects study were not yet determined at the time of the plant design, equipment was not included. Thus, this space was constructed and left relatively empty (a few laboratory cabinets and one bench were included).

## Public Information

Increasing public awareness about potable reuse was one of the main objectives of the Project so several features were included in the plant design to help achieve this goal. Earlier public opinion polls had confirmed that plant tours would have the most impact on public perception. Enhancing the public tours, thus, became a focus of the plant design. As part of this effort, the plant effluent (product water) initially was to be displayed in an on-site lake surrounded by an attractive park. During the early stage of design process this concept was questioned. The Project advisory committee recommended strongly against degrading the effluent quality in an outdoor reservoir. The committee also felt very strongly that recreation in this particular location would suffer from the association from odors and appearance of the surrounding industrial area. The on-site lake and recreational facilities, thus, were deleted from the design. Instead, the effluent was displayed inside the plant under controlled conditions using a stainless steel water fountain sculpture. The bulk of the plant product water, after a small amount was removed for testing, was used for Reuse Plant landscape irrigation and process water at the Metropolitan Denver Sewage Disposal District.

Many other features were designed into the plant to facilitate tours. The Demonstration Plant was located close to York Street a major collector street in Commerce City. Access was available from Interstate 25 and downtown Denver along easily traveled streets. The plant, associated processes, parking, and landscaping occupied approximately ten acres of the twenty acre site in proximity to this street. Except for the lime clarification system, all processes were housed indoors. Landscaping was used to isolate the facility and the parking lot from the surrounding industrial blight by berms and plantings.

The main process building itself was a metal sandwich construction with an attractive glass entryway and a five-story 40,000 ft.² main process area. Approximately one-third of the main process building was devoted to administrative functions including laboratory, sample preparation area, side stream testing room, offices, reception area, and tour gathering facilities. The entire administrative wing was attractively furnished with appropriate portions carpeted and decorated. The tour gathering area consisted of a carpeted waiting area and conference room with audiovisual capabilities for seating large tours (about 80 person capacity). It doubled as a conference room and classroom for plant staff. In the tour gathering area inside the entryway were displays for visitors access to color graphics displays with limited hands-on capability.

Wide isles were provided to accommodate relatively large tours and all public areas were handicapped accessible. At strategic locations throughout the plant displays were located to explain nearby process functions. Some of the displays included hands-on opportunities and were animated. All contactors, reaction tanks, piping, valves, and other vessels in the main process area were color-coded to assist with tours. The tour route was laid out to follow the path of water through the plant.

Viewing windows provided guests the opportunity to see into the control room, computer area, and laboratory in the administrative wing without disturbing workers inside. The control room had a massive L-shaped consul with hundreds of gauges and adjustment knobs. The annunciator alarm panel hung from the ceiling displaying lit alarm conditions and sounding an audible alert to operators. Visitors looked into the room from behind a glass wall and could see operators at their computer terminals attending to plant operation. The whole scene was reminiscent of nuclear power plant control rooms seen in movies of the day.

In the main plant process area some processes were contained in separate rooms (carbon regeneration furnace, ozone, chlorine dioxide, and reverse osmosis high pressure pumps). Large windows were installed in each of these rooms so that tours could see what was inside, thus, assisting the tour guides in their efforts to explain the function of each process.

~ ~

The Reuse Treatment Plant design completion was signaled by the construction bid award in December 1980. The bids were based on the construction bid documents that incorporated a series of treatment processes and equipment needed to convert unchlorinated secondary wastewater effluent to drinking water. The plant was designed for continuous operation at the maximum influent flow of 1 mgd. Also, included were the laboratory facilities and on-line data acquisition and analysis systems to acquire all the operational and scientific data needed to evaluate the plant performance. Features to facilitate tours included color coded piping and equipment, informational displays, audio visual capabilities, and decorated areas that would host large groups. The plant appearance from the outside was enhanced by landscaping that served to screen visitors from the surrounding industrial area. The result of the Reuse Treatment Plant design was a facility that was later described as "the world's most complex water treatment plant."

# 4

# Plant Construction

The *Notice to Proceed* with construction was issued on Thursday, January 29, 1981. Work started at the site on Monday, February 9 with excavation of the ballast ponds, regeneration basins and mechanical basement for the main process building. The contract completion date was January 29, 1983 exactly two years from the contract initiation. Problems with weather, a fire at the construction site, and labor problems added more than 300 days to the initial construction schedule. Even with the authorized contract extension plant construction should have been completed by the end of 1983 but it wasn't fully functional until October, 1985.

As the scheduled construction completion date approached the general contractor and subcontractors developed serious disagreements. Significant delays then resulted to complete the warrantee and construction defect repairs. The completion of the reverse osmosis system, ozone process, and the activated carbon regeneration furnace were severely affected by these issues. The Water Department had to complete many of the remaining items and perform much of the warrantee repairs before the plant became be fully functional. The plant process optimization phases were delayed almost two years. Even after the formal start of the process assessment period the ion exchange system couldn't be operated due to construction deficiencies that still needed repair.

One example of an item where the Water Department had to intervene was the reverse osmosis system. The equipment supplier had a serious concern about the warrantee requirements included in the construction contract. Due to this dispute the supplier delayed delivery of the system. The problem basically involved certain guarantees about the performance of the membranes. A reverse osmosis system change order that resolved this issue was finally approved after more than three years of negotiations. The Water Department agreed to modify the specification paragraphs in question for several considerations: the owner received a no cost service agreement for one year and one complete reload (126 modules) of reverse osmosis membranes.

## Plant Operations Staffing

Before beginning operational testing, staffing plans for the Project were developed in 1981. These plans were discussed with the Water Department management as well as the Project advisory committee. Contrary to the recommendations of the reuse staff and the Project advisory committee, Water Department executive management decided to structure the Project as a matrix type organization. This structure resulted in Project employees coming from two Divisions and several Sections. Most operating and water quality analysis Project personnel, then, technically reported to their respective Departments instead of directly to the Reuse Project Manager.

Following this decision the Water Department hired a Reuse Plant Supervisor in December 1982 so that he could become familiar with the plant and the start the hiring process for the remaining personnel. All lead operators were hired by May 1983 because the contractor had scheduled start-up for June 1983. Equipment training was to occur between May and early June. Training was conducted from May through August and initial startup did not occur until January 1984. This so called "startup" was a construction contract required seven-day performance test of the plant and equipment using Denver drinking water for the plant influent. The result of this test was to identify any defects that would then need repair.

The seven-day performance test was delayed six months so the operations staff used this time to gain a strong understanding of the plant and processes. The staff prepared and conducted training sessions on the various unit processes. These training sessions were filmed and incorporated with the equipment supplier training films to provide a library of operations training videotapes. The operators also wrote an operation and maintenance manual by modifying the draft manual provided by the design engineering firm as part of the services during construction. These activities produced a highly trained and qualified operating staff that was responsible for the operation of the complex treatment plant and to participate in the optimization of the multitude of process systems.

## Construction Defects and Repairs

Ahead of the contractor required seven-day plant reliability test the Water Department purchased treatment chemicals in October and November 1983. The plant operational test was then conducted using drinking water for the plant flow January 4-13, 1984. All treatment plant processes were run during the test except ozone, reverse osmosis, and the carbon regeneration furnace. Plant operation was performed by the Water Department personnel with the concurrence of the contractor. Contractor personnel were on 24 hour-a-day call during the test. Although several problems were encountered, the test was successfully completed on January 13, 1984.

Following the completion of the seven-day test using drinking water for the plant flow, secondary treated wastewater was brought into the plant for the first time on January 16. Treatment was begun through two-stage lime, recarbonation, filtration, and chlorine dioxide disinfection. All other treatment processes were bypassed initially. A phased operations approach was adopted to bring each new process online as previous process sequences were stabilized. Start-up and optimization of the treatment plant processes continued through September 30, 1985. This protracted start-up was caused by the incomplete construction defect repairs and testing of several of the water treatment processes which had to be completed by the Water Department.

Shortly after starting continuous plant operation using wastewater numerous mechanical, electrical, instrumentation, computer system, and chemical system problems surfaced. Many of these problems were related to various construction defects or omissions as well as equipment flaws that required warrantee repairs. As a result of serious disagreements between the contractor and the subcontractors, timely correction of these problems was not possible. Many months passed without action on critical process components. This situation necessitated bypass of some primary treatment processes and inadequate operation of other processes.

No unit process was spared from the various construction defects or omissions and warrantee repair difficulties. Notable examples of the most important treatment process defects illustrate the scope and impact these issues had on the Project.

The chemical treatment DC variable speed mixers began failing shortly after the seven-day performance test. Speed control units on the mixer drives were installed in an area where they were subjected to temperature extremes beyond conditions specified by the manufacturer. Additionally, operation at high altitude required different choices for the mixer motors. To correct this crucial aspect, the motors were replaced with larger units and new speed control units were installed in the raw water meter vault where temperature was controlled.

Filtration was affected by ineffective flow control valve operation. The valves supplied under the contract were a hybrid configuration with equipment from four different suppliers. Excessive leaking, and extreme instrument and mechanical issues finally required replacement with a single supplier unit. The flow control valves in the ion exchange columns were replaced for the same reasons. The replacement valves performed well for the duration of the Project.

The ion exchange system was operated for only one week during the first quarter of 1984 when it became apparent that the Ammonia Removal and Recovery Process (ARRP) tower fans were not functioning properly. Following investigation it was determined that they have been installed improperly and were rotating in the wrong direction. The direction was reversed and performance improved, but then the fans developed extreme noise and vibration. New fan belts were installed, fan bearing bearings were replaced, and motor bearings were replaced with no effect. These fans were critical for the process, so operation was terminated until they could be repaired.

Months passed without any progress. The contractor didn't proceed in a timely manner to solve this problem; as a result the Water Department was again forced to make the necessary repairs. The fans finally were replaced more than a year and a half following the identification of the problem. By then it was the third-quarter 1985 and the plant optimization operation phase 1 startup was then imminent. There was not enough time to optimize this complex process so it had to be bypassed. During phase 1 operation laboratory-scale tests were performed to gain a better understanding of the system characteristics before its eventual start-up in phase 2.

First and second-stage carbon columns were inspected towards the end of the first year of operation while the plant wasn't yet fully functional. Previous difficulties experienced during the application of corrosion protection lining suggested that it would be prudent to examine for evidence of corrosion. It was determined that the stainless steel well screens had been installed as specified with epoxy coated bolts but with *mild steel washers* which had all but disappeared. The bolts themselves had holes in the coating and were also corroding. The bolts and washers were replaced with stainless steel hardware. The urethane lining had survived reasonably intact except for the hatchways. Evidently the urethane had not been applied beneath the cover seat and water had penetrated between O-ring and hatch. Corrosion was so severe that one hatch required dismantling and rebuilding.

The ozone system was tested in May 1984. The performance test was successfully completed with only one of the generators in operation. The second ozone generator was performance tested and placed in service later in 1984. After that the complete system operated consistently.

The original chlorine dioxide system was oversized to the extent that reliable control was not possible. A second generator was provided by the sodium chlorite supplier. This generator was operated from June 1984. A third modified generator capable of achieving higher conversion efficiencies was then installed and this operated successfully thereafter.

The plant computer system, which was necessary to assist operators and monitoring the multitude of processes and systems, failed its 30-day reliability test during the second quarter of 1984. Settlement was reached to avoid a repeat of the test. Several enhancements, agreed upon jointly by all parties, were added because of the settlement and these made the computer system a much more valued operational tool. Operations staff relied heavily on the computer monitoring system and it was a key tool for performance evaluation and water quality assessments.

The reverse osmosis system was not brought online until more than the year following the seven-day test. Vibration in the high pressure pumps caused a major delay. New pump bases were fabricated and installed to resolve the situation. The reverse osmosis system was operated beginning the first quarter of 1985, following the repair of numerous control and mechanical defects, the 30-day performance test finally was completed and the system was operated consistently after that.

The last major process to begin operation was the activated carbon regeneration furnace. Functional testing was completed in 1984 but the required performance test was not started until October 1985 due to problems between the system supplier and the general contractor. A separate arrangement with the Water Department was necessary to secure startup services from the supplier. The startup test did not occur until the first quarter of 1986. This test revealed several more deficiencies that needed repair before the system was functional. System operation was initiated after these adjustments later in 1986.

## Summary

The Reuse Plant was operated inconsistently from January 1984, after completion of the seven-day reliability test, until the phase 1 testing began on October 1, 1985. This extended period of incomplete treatment was primarily due to complications arising out of the contractual conflict that prevented the completion of work to correct construction defects and warrantee repairs. This caused a significant delay for full plant startup. In addition, plant operations and maintenance staff had to be utilized for construction defect and warrantee repair work to complete these tasks as quickly as possible. Even though numerous mechanical, electrical, and control system problems plagued the various systems through construction, much was learned about the operation of the complex processes. Valuable data were gathered to establish standard operating ranges that were used in the process evaluation testing phases.

# 5

# Experimental Design

A Project experimental design report was required by the US EPA Cooperative Agreement. This document described the methods used to conduct the scientific and engineering programs that were needed to satisfy the Project goals. These major programs included: animal health effects studies, water quality studies, and plant operations.

The whole-animal lifetime health effects study was a unique element of the Project. There were many unknowns about the sample preparation and the testing protocol that needed to be defined since this comprehensive study had not previously been performed on drinking water. Thankfully, there were two and a half years of process evaluation to be completed before the scheduled start of this part of the program. The planning for the animal health effects study then was placed on a separate track with its own expert advisors and timelines. Discussions of these developments are the topic of Chapter 12 and these were not included in the Project experimental design report.

## Quality Assurance

All the scientific and engineering programs included extensive quality assurance procedures to ensure the credibility of the data used as the basis for the Project conclusions. An elaborate series of checks, safeguards, and reviews were established to ensure the integrity of the data used for water quality comparisons and the plant operational evaluations. The two largest quality assurance programs covered the water quality studies and the plant operation.

The water quality studies quality assurance manual (14 volumes) addressed the topics:

- Laboratory equipment and environment
- Sample preservation
- Sample shipping
- Chain of custody
- Administration
- Inorganic Chemistry

- Organic Chemistry

- Virology

- Coliphage

- Microbiology

- Contract Laboratory's

Quality assurance data was collected and reviewed to track accuracy and precision of each analysis method. The manual was constructed so it could be used to satisfy the demands of the State as well as US EPA. It was prepared in loose leaf document style to facilitate changes and updates. This was a working document designed to be used on a daily basis rather than to satisfy a formal requirement and then set aside.

The reuse treatment plant operation had its own quality assurance manual (11 volumes) that contained the procedures for these major topics:

- Equipment identification

- Equipment accuracy and precision

- Calibration procedures

- Frequency for verification checks

- National Bureau of Standards traceable standards and testing

- Secondary standards testing and adjustment frequency

- Records

- Acceptable deviation and action points

These procedures were applied to all measuring devices and instruments including: flow meters, pressure sensors, temperature, level, chemical metering pumps, on-line chemical sensors, UV intensity meters, and various process instruments. In all about 500 measuring devices and instruments required quality assurance procedures.


## Data Handling and Evaluation Methods

Data for the Project was obtained from input from over 800 real-time plant sensors, laboratory test results, operational records, managerial information, public information program, outside agency data, and maintenance records. The vast majority of this information was collected in the plant computer system. The system a PDP 11/44 Digital Computer® system was designed to acquire all the real-time data from the plant as well as the results of laboratory tests and certain other operational data. The data was then displayed for use by plant operators for plant control or archived for later recall. Laboratory data was entered manually at the Reuse Plant site and remotely at the Water Department Quality Control Laboratory and then was available in a variety of reports designed to show correlations and key statistical relationships.

Some items were not suitable for entry on the main Project computer and were better handled on a personal computer system (new at the time). This system was used for operational needs such as producing pump curves and to incorporate data and graphics together for report generation.

### Project Statistics

The statistics used for data analysis followed those developed by Dr. Perry McCarty, professor of environmental engineering at Stanford University, for his highly respected work at Water Factory 21 and M.C. Kavanaugh, J.M. Montgomery Engineering, who directed the Potomac Estuary Experimental Plant for the Army Corps of Engineers. These projects, like the Reuse Demonstration Project, involved the treatment of wastewater by sophisticated processes. Data from environmental samples were, thus, encountered and required statistical analysis.

The approach they developed, and was being considered by the Reuse Demonstration Project, was evaluated by the Denver Water Department consultant, Dr. Joel Cohen, professor of mathematics at Denver University. Lognormal statistics were used to describe the central tendency and data variability. Dr. Cohen developed a mathematical approach that was adopted and critically reviewed to deal with the situation where part of the data was effectively censored from view since some of the data was below detectable levels. Also, a method for determining statistically different geometric mean estimates between two data sets was defined. Dr. Cohen wrote two articles describing and critically reviewing these statistics and these descriptions formed the basis for the eventual adoption of the suggested statistical methods by the Project Advisory Committee.

The statistical approach proved useful when making the comparisons required to satisfy the objectives of the Project. The statistics were periodically reviewed to verify that the methods selected were presenting an accurate picture of the true sample populations that were encountered.

## Water Quality Studies

The water quality studies program (analytical studies) was designed to provide sufficient water quality information to support the Project objectives. These included establishing water safety and evaluating individual and aggregate process contaminant removal performance. Important elements of the program were needed to achieve these objectives such as: establishing a routine sampling and analysis schedule needed to determine the product water safety in comparison to Denver drinking water; developing sampling and testing strategies to enable evaluation of unit process performance and establish contaminant removal capabilities; and, the procedures to conduct the plant-scale contaminant challenge dosing studies. This last study was performed before beginning the animal health effects study. It involved adding selected contaminants to the plant influent at high levels to determine the removal capability of this unique treatment facility.

The reuse water quality studies program listed the parameters proposed for the testing and the frequency and types of samples that would be used. Work on this program was begun by first dividing it into main two major phases. The design phase which included the development of the routine sampling and analysis program, and the operation phase that ran concurrently with the demonstration plant operation using the program developed during the design phase.

In general, the approach utilized the best available methodology to accomplish the various evaluations. Recognized methods were used when recommended by the expert Analytical Advisory Committee members. Most test methods were approved and included in the industry reference *Standard Methods of Analysis for the Examination of Water and Wastewater*. A few, though, were new and not yet officially approved but the expert advisors strongly supported using these methods that were later approved. Basic methods development research was not conducted. The focus of the Project was applied process oriented research.

During the three-year period of the demonstration plant construction many questions about the routine sampling and analysis program were answered. Since most water quality testing was performed by the Department's own laboratory, a great many adjustments had to be made internally to change the emphasis from routine water monitoring to analysis methods that extended the limits of detection and emphasized contaminant identification. Laboratory personnel were added and trained, new equipment purchased, and the laboratory expanded to accommodate the number and variety of samples that were required for the evaluation of the reuse treatment process.

A preliminary version of the proposed water quality studies program was presented to the analytical and health effects advisory committee at its first meeting (1981). Changes were then made and incorporated into the modified water quality testing program. One of the first items to be determined was the parameters to be examined. In addition to all the regulated parameters any substance that might be present in treated wastewater was included. Virtually every possible water contaminant became part of the Reuse Project water quality testing program (Table 5-1, 11-2, and 11-3). Thus, the water quality testing program included many substances never routinely tested in conventional water supplies (e.g. rare earth elements, radioactive isotopes, unregulated trace organic chemicals, particle count, and many other unregulated parameters).

Testing was not only designed evaluate quality of the reuse product water compared national water quality standards but also to Denver's drinking water supply. Denver's drinking water was selected as a comparison standard for two reasons: (1) public opinion surveys had revealed that a majority of Denver residents would accept reuse water only if the quality were equal to or better than the present supply; (2) there were no recognized standards which applied to reuse water. Denver's drinking water (which was derived from a relatively protected and unpolluted source) met all health standards and there was no reason to believe it would ever fail to satisfy future health standards. Duplicating this water thus provided a margin of safety between existing and future standards.

Another major item to be addressed was the frequency and type of samples for a routine sampling program. This was necessary to facilitate decisions concerning testing equipment and laboratory personnel needs. The routine sampling program was designed to provide enough information to define the frequency distribution of concentrations (central tendency and variability) in the plant influent, plant effluent, and Denver's potable supply. Also, the test data was used to evaluate the contaminant removal efficiencies of the intermediate treatment processes.

Twenty-four hour composite samples taken every six days was adopted for the primary sample locations. The six day sampling frequency was selected to obtain samples on different days of the week. This resulted in both weekly and daily variations being observed. The water quality program was scheduled for five years so sufficient data was obtained to evaluate the plant performance on an annual basis and provide a very substantial database.

Sample frequency was prioritized based on importance, potential health impact, presence in plant influent, and test method complexity. Parameters tested less frequently still provided sufficient data over the life of the Project to make the necessary evaluations. For the most part, these parameters also provided information that supported other analyses. For example, bacterial pathogens supported bacterial indicator tests such as coliform and standard plate count. Even though some pathogens weren't analyzed often enough to provide annual statistical evaluations as significant as those sampled more frequently, over the five year life of the Project sufficient data was gathered to make the necessary evaluations.

A secondary goal of the analytical studies program was to determine the contaminant removal efficiencies of the individual plant processes. To accomplish this, special paired samples were taken before and after each major process. The samples were taken at the same time as regular sampling for the influent and effluent. For example, trace metal analyses were performed weekly for the effluents and every third week for other sample locations.

This routine sampling program (Table 5-1) represented the minimum number of analyses that were performed at the Reuse Demonstration Plant. These results often signaled the need to conduct special tests. For example, hourly grab samples around certain processes were necessary to augment composite samples to explore the effects of possible daily fluctuations. Also, special samples of the many process waste streams were required. Additionally, the routine program was designed to allow analysts enough time to explore special studies in addition to the routine analyses.

| Analyses | Raw | Lime | Filter | UV | Carbon | RO | Air Stripping | Ozone | Plant Eff | UF | Denver Tap |
|---|---|---|---|---|---|---|---|---|---|---|---|
| Process Performance[1] | WG | WG | WG | WG | WG | WG | WG | WG | WG | WG | WG |
| Control Tests[2] | C/D | C/D | C/D | C/D | C/D | C/D | C/D | C/D | C/D | C/D | |
| Bac-T Indicators[3] | DG | DG | DG | DG | DG | DG | DG | DG | DG | DG | DG |
| Fecal Coliform | WG | | | | WG | | | | WG | WG | WG |
| Fecal Strep | WG | | | | WG | | | | WG | WG | WG |
| Bac-T Pathogens[4] | QG | | | | QG | | | | QG | QG | QG |
| Microscopic Examination[5] | QG | | | | QG | | | | QG | QG | QG |
| Coliphage | WG | WG | WG | WG | WG | WG | WG | WG | WG | WG | WG |
| Enteric Virus | | | | | | | | | MG | MG | |
| Metals | WC | | | | WC | | | | WC | WC | WC |
| Radioactive Screening[12] | WC | | | | WC | | | | WC | WC | WC |
| Inorganic and General[11] | WC | | | | WC | | | | WC | WC | WC |
| Rare Earth Elements[6] | QC | | | | QC | | | | QC | QC | QC |
| Radioactive Isotopes[7] | QC | | | | QC | | | | QC | QC | QC |
| Cyanide | QC | | | | QC | | | | QC | QC | QC |
| Asbestos | QC | | | | QC | | | | QC | QC | QC |
| Purgeable Organics[8] | EC | | | | EC | | | | EC | EC | EC |
| Contract Organics[9] | QC | | | | QC | | | | QC | QC | QC |
| TOX[10] | MG | | | | MG | | | | MG | MG | MG |

## Footnotes

[1] TOC, hardness, alkalinity, turbidity, ammonia, specific conductance, total coliform, coliphage, mHPC

[2] Turbidity, temperature, dissolved oxygen, pH, nitrite, chlorine residual, specific conductance, ozone residual

[3] Total coliform, mHPC

[4] *Salmonella, Shigella, Campylobacter, Clostridium perfringens, Legionella*

[5] *Giardia, Cryptospordium, Entamoeba histolytica, Endamoeba coli,* Helminths, Nematodes, Algae

[6] entire list of rare earth elements by spark source mass spectroscopy

[7] Radon, plutonium, tritium, radium 226, radium 228

[8] Volatile organic compounds (EPA 502.2), Grob Closed Loop Stripping GC/MS (EPA 8270)

[9] Pesticides, herbicides, carbamate pesticides, acid extractables, disinfection by-products, haloacetic acids

[10] Total organic halogen

[11] Sulfate, nitrate, nitrite, phosphate, chloride, fluoride, bromide, MBAS, TDS, TSS, TKN, ammonia, odor, color, conductivity, alkalinity, hardness, TOC, silica, particle count

[12] Gross alpha activity, gross beta activity

## Sample Frequency Codes

| | |
|---|---|
| C | continuous |
| D | daily grab |
| WG | weekly grab |
| MG | monthly grab |
| QG | Quarterly grab |
| WC | 24 hr composite taken every 6 days |
| EC | 24 hr composite taken every 18 days |
| QC | 24 hr composite taken every quarter |

## Plant Operations

The plant operation program defined the staffing, training, operation, maintenance, process evaluations, plant data storage and retrieval, process data quality assurance, and cost assessment elements needed to achieve the Project goals. These functions, often performed without recognition, were critical to the plant function and consumed considerable labor and resources.

Maintenance records for all the plant instrumentation and mechanical equipment were kept in a PC based, commercial maintenance management program. The program stored nameplate data, maintenance procedures and a parts inventory for each piece of plant equipment. This included information for more than 80 pumps, 100 motors, and 500 instruments. This program was used to schedule preventative and predictive maintenance activities, and held maintenance history for all the plant equipment.

One of the major Project goals was to provide data for full-scale implementation. Data from plant operations records was needed to develop accurate cost estimates for this purpose. Although construction and operating estimates could and were made using standard engineering tables, the estimates obtained from actual operating experience were far better. Hence, data was continuously collected from the Reuse Plant operation for power usage, chemical usage, and labor. This data, then, was substituted for engineering table values to provide costs that were based on actual operating experience.

### Project Staffing

The Reuse Demonstration Project organization (Figure 5-2) was designed to accomplish the tasks necessary to achieve the Project goals by providing adequate qualified staff and necessary professional support. The numbers shown in Figure 5-2 for Water Quality testing (15) were not exclusive to the Reuse Project. Most analysts split their time between Reuse Project samples and routine Denver drinking water quality control samples. The number shown below the Reuse Project Manager (5) includes personnel that were added to prepare samples for the animal health effects study.

The bulk of the Project staff was involved in plant operations. The experimental design called for the daily operation of the Reuse Plant to be conducted by two operators assigned to a 12-hour shift that was used at Denver's water treatment plants. A Project objective was to demonstrate that the highly complex reuse treatment plant could be operated with a similar number and qualification as operators at Denver's conventional water treatment plants.

The lead operator technician at the Reuse Plant was required to be certified by the State of Colorado at the "A" (highest) level in either the water or wastewater classification and at the "D" level in the other discipline. The assistant technician, on a two-person shift, was required to be certified at the "C" level in either discipline and the "D" level in the second. Many of the technicians achieved certification levels above those required. The State of Colorado also offers a certification in industrial classification for personnel required to operate these types of processes, many of which were used in the Reuse Plant. At one point during the Project seven members of the operating staff were certified at the "A" level in all three classifications.

The operators (technicians) were responsible for process monitoring and control. Technicians normally assigned to maintenance duties replaced the plant operators in their absence. These fill-in personnel met the same classification requirements as the normal shift technicians.

The training of the operations and maintenance staff was done by Project personnel. Each technician was trained in plant operations through a series of classroom sessions which included videos and training manuals. The manuals and videos were produced by the operations and maintenance staff at the beginning of the Project. Each of the initial plant technicians were assigned a unit process and were responsible for preparing an operations manual and video for that process. The information covered in the manuals and tapes ranged from textbook information on the treatment to information specific to the reuse facility. In preparing the training materials the technician became an in-house expert on the process. Videos were also made when manufacturer representatives were on-site to start up equipment they supplied. The combination of the training materials and the on-site experts proved to be the valuable resources during the operations phases of the Project.

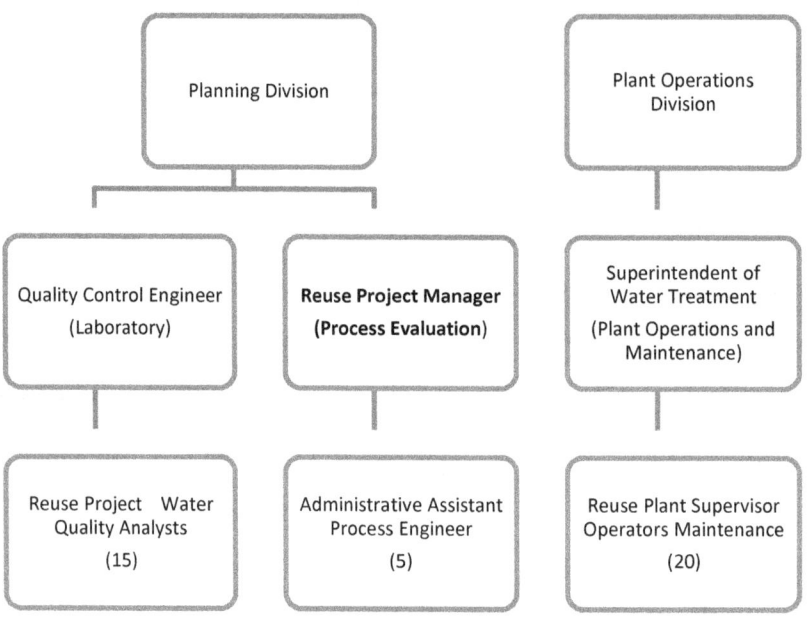

**Figure 5-2: Reuse Project Organization**

## Summary

The Reuse Project experimental design described the procedures that were used to test and evaluate the scientific and engineering studies. Quality assurance was an emphasis to establish the credibility of the data used for assessing the achievement of the Project goals. Data handling and statistical procedures were critically reviewed to ensure the best methods were used. Plant maintenance and instrument repair procedures were defined and carefully recorded for use in process evaluation and for developing accurate treatment costs. Project personnel received specific training necessary to accomplish the complex requirements of this unique Project. Thus, the experimental design defined the Project components critical to a successful outcome.

# 6

# Process Assessment

Following the extended construction period, that included inefficient plant operation for about twenty-one months (January 1984- October 1985), a formal treatment process assessment program was initiated. The two and a half-year program (October 1985-March 1988) was designed to evaluate all the processes included in the demonstration plant. Information from these treatment process assessments were then used to select the most effective and cost efficient treatment sequence to use during the two-year animal health effects study period. An additional three-month plant operational period (October 1988 - January 1989) using the health effects study treatment sequence was used as a shake-down period before that crucial study began and during which the planned contaminant challenge studies were conducted.

The plant was operated in several treatment configurations, identified as phases, during this assessment period. Four phases lasting 4 to 11 months each were used to operate the treatment plant, optimize process performance, and evaluate the reliability and effectiveness of each treatment sequence.

The dates of operation, a process sequence descriptor, and a list of primary treatment processes are listed for each operational phase.

***Phase 1*** *(October 1985-March 1886)*

The entire reuse treatment plant was operated except for the selective ion exchange process. The treatment processes included: Lime Clarification, Recarbonation, Filtration, First-stage Activated Carbon Adsorption, Ozonation, Second-stage Activated Carbon Adsorption, Reverse Osmosis, Air Stripping, and Chlorine Dioxide Disinfection.

***Phase 2*** *(April 1986-October 1986)*

The entire reuse treatment plant, as designed, was operated. This included: Lime Clarification, Recarbonation, Filtration, Ion Exchange, First-stage Activated Carbon Adsorption, Ozonation, Second-stage Activated Carbon Adsorption, Reverse Osmosis, Air Stripping, and Chlorine Dioxide Disinfection.

## Phase 3 *(November 1986-February 1987)*

The reuse treatment plant was operated without reverse osmosis and air stripping. The treatment processes included: Lime Clarification, Recarbonation, Chlorine Dioxide Disinfection, Filtration, Ion Exchange, Ozonation, Second-stage Activated Carbon Adsorption, and Chlorine Dioxide Disinfection.

## Phase 4 *(April 1987-March 1988)*

Several treatment process sequences were operated in a series of mini-studies. These studies provided information needed to make a decision about the processes that were used during the animal health effects study. At the end of this phase the health effects treatment sequence was chosen and operated to optimize performance before starting the animal health effects study. The health effects study processes included: Lime Clarification, Recarbonation, Filtration, Ultraviolet Irradiation, Second-stage Activated Carbon Adsorption, Reverse Osmosis, Air Stripping, Ozonation, and Monochloramine Disinfection; and a pilot-scale Ultrafiltration system.

# Treatment Process Operational Experience and Assessment

The processes designed into the reuse treatment plant were all tested during the assessment phases (Phase 1, 2, 3, and 4) of the Project. The operational experiences and performance results for each treatment process illustrate the many modifications that were needed to achieve optimum system function.

## Lime Clarification (single and two-stage modes)

The lime treatment and recarbonation systems were operated during all the phases of plant operation. Several modes of operation were evaluated before selecting the final treatment configuration. These modes ranged from ferric chloride instead of lime as the primary coagulant to two-stage softening and recarbonation. Single-stage lime clarification was chosen as the treatment process that was used for the duration of the health effects study program.

The pH setpoint of the rapid mixed basin was varied to determine the effect on water quality. The minimum solubility pH for calcium carbonate occurs at 9.5. For softening purposes this was the optimum pH setpoint. However, for potable reuse treatment this pH was too low to benefit from virus inactivation and removal of numerous potential contaminants obtained at higher pH levels. The process pH setpoint of between 10.5 and 11.5 received further evaluation. The advantage of using a setpoint at a lower end of this range was reduced chemical usage and less sludge production. The disadvantage was that the clarifier effluent turbidity levels increased to unacceptable levels at the lower end of the range. The best value for these reasons was found to be an intermediate pH of 11.0.

Sodium hydroxide was examined as a substitute for hydrated lime. This treatment achieved high pH required for metals removal, bacteria, and virus inactivation while keeping sludge production low. The ease of feeding liquid sodium hydroxide also made this an attractive alternative. The two major disadvantages of using this treatment option where the cost and increased alkalinity. Sodium hydroxide costs approximately twice as much as hydrated lime to achieve the same pH. The higher conductivities and alkalinity produced from sodium hydroxide use created higher operating pressures in the reverse osmosis system. The increased alkalinity also caused an acid demand in the reverse osmosis feed water that exceeded the pumping capacity of the chemical feed pumps.

A combination of lime and sodium hydroxide was used in full-scale plant tests. This was accomplished by feeding a baseline dosage with lime and using the closed loop pH control system to trim the sodium hydroxide to meet the pH setpoint. The goal of this trial was to reduce the large swings in pH in the rapid mix basin when using lime alone. This goal was accomplished but at the expense of creating operational problems like those seen when sodium hydroxide was used alone. A large and unacceptable increase in operating pressure within the reverse osmosis system was one consequence. Sodium hydroxide was, necessarily, rejected for use in the health effects study treatment sequence.

Two-stage softening and recarbonation was the first process configuration used following the plant startup. The pH setpoints were 11.5 in the first-stage and 9.5 in the second-stage. Carbon dioxide was used for pH adjustment. Soda ash was added as a carbonate source to the second-stage rapid mixed basin. The hardness in the effluent was controlled by varying the soda ash dosage. This treatment provided excellent water quality but created unacceptable operational problems. The physical layout of the rapid mix and recarbonation basins required pumping the flow from the first-stage to the second-stage. At a pH of 9.5 this water was unstable and deposition of calcium carbonate within the pumps was so severe that acid cleaning was required on a weekly basis. This frequency and practice was unacceptable since it affected the reliability and the cost of maintenance for this process so two-stage lime softening treatment was discontinued.

Ferric chloride was also evaluated as a primary coagulant alternative to lime. After about two weeks in operation it was obvious that this was not acceptable. The turbidity of the clarifier effluent was very high while the relatively low pH, compared to lime, did little to remove metals, reduce bacteria, or inactivate viruses. Ferric chloride was not an acceptable substitute for lime clarification.

One of the primary goals of the coagulation process was to provide a consistent water quality to the downstream processes, especially the reverse osmosis system. Of particular concern were calcium, silica, sulfate, total organic carbon, and bacteria. The lime treatment offered superior results for all these pollutants over ferric chloride. Virus inactivation from lime treatment was important to produce biologically safe water. The selection of lime treatment was made at the completion of the preliminary trial modes of operation. Due to maintenance problems created using two-stage lime treatment, single-stage operation was the treatment of choice for the health effects study.

During the first four phases of operation several coagulant aids were tested. Polymer coagulant aids were added in both the rapid mix and flocculation basins. None of the polymers tried provided a significant improvement in water quality and there use was discontinued.

Ferric chloride was also used as a coagulant aid. Jar tests showed that the feed point of the ferric chloride made a significant difference in its effectiveness. Plant scale testing included an evaluation of the ferric chloride addition point. After testing several feed locations, the overflow weir of the rapid mix basin was found to yield the best results. The existing polymer feed lines at this location were converted to ferric chloride and this feed point was used for the remainder of the plant operation.

The optimum settings for the lime clarification process was a single-stage treatment with a pH setpoint 11.0, ferric chloride was used as a coagulant aid at a dosage of 20-25 mg/L. This dosage was adjusted as needed to maintain the clarifier effluent turbidity.


## Recarbonation (carbon dioxide or sulfuric acid)

The recarbonation system was designed to feed carbon dioxide gas into the process flow stream. Strong acid neutralization with sulfuric acid was provided as a backup. The volume of carbon dioxide required for single-stage neutralization was underestimated in the original design. As a consequence carbon dioxide feed system was inadequate. To solve this problem a baseline dosage of sulfuric acid was fed to the flow stream and carbon dioxide was added to meet the pH setpoint. Changing components within the carbon dioxide flow control valves and pressure reducing valves increased the feed capacity and eliminated this problem. Sulfuric acid then was only needed as an emergency backup. One problem with sulfuric acid was the increase in the sulfate content of the product water. This created a potential for irreversible calcium sulfate fouling in the reverse osmosis system.

Closed loop control of the pH in the recarbonation basin was reliable and problems were minimal. The pH setpoint for recarbonation was originally set at 7.2. This accomplished neutralization of the water but created an operational problem downstream as water flowed through the ballast ponds that followed of the recarbonation basins. Aeration in the ballast ponds stripped dissolved carbon dioxide from the process flow and upset the chemical equilibrium established in the recarbonation basin. This resulted in severe scaling in the internal components of the filter supply pumps that used the ballast ponds for pump suction. The problem was eliminated by increasing the pH setpoint of the recarbonation basin effluent from 7.2 to 7.8. At the higher pH dissolved carbon dioxide was not stripped by the ballast pond aeration system.

## Pressure Filtration (filter aid)

The filtration process was brought online during initial plant start-up in conjunction with the lime treatment system and operated continuously thereafter. After overcoming some deficiencies in the installation of the media, the system performed well. A study that investigated polymers as filter aids was conducted during phase 3 operations. None of the polymers investigated were found to be effective.

One of the operational peculiarities that developed in the filtration system was a large drop in dissolved oxygen across the filter bed. Initially it was thought that this was due, in part to air binding within the filter. Upon further investigation the loss was found to be due to bacteria that had colonized the media. Aeration in the ballast pond ahead of the filters successfully added high levels of dissolved oxygen. This practice though promoted growth of nitrifying bacteria in the filters. The problem was mitigated by measuring nitrite concentrations in the effluent of each filter on a weekly basis. This problem was eliminated by disinfecting the filters while offline with chlorine dioxide if the nitrite level exceeded 0.4 mg/L (about a monthly occurrence). The filters were routinely operated until the head lost reached 11 feet at which point a backwash was initiated.

## Selective Ion Exchange

The primary ammonia removal barrier incorporated into the demonstration plant design was selective ion exchange. The naturally occurring zeolite, *clinoptilolite,* is selective for the ammonium ion in preference to all other cations other than potassium. The regeneration of the zeolite was accomplished with a sodium chloride brine solution. To conserve brine, and thereby minimize plant waste stream, the brine was processed through a regenerate recovery system. This system, known as the ammonia removal and recovery process (ARRP), removed ammonia gas from the regenerant brine solution at high pH followed by absorption with a sulfuric acid solution. The resulting solution of ammonium sulfate was then collected and stored for use as a fertilizer.

The ion exchange system went through a prolonged startup. Beginning in 1984 several design and construction related deficiencies limited ion exchange performance. Leaking valves resulted in incomplete regeneration of the brine and this affected the ammonia removal results. Excessive vibration in the ARRP fans delayed system optimization. Sodium hydroxide was used to prepare the spent regenerant for air stripping. This resulted in the immediate caustic softening of the brine, scaling the downstream mixer and influent piping. The chemical addition point was subsequently relocated. Once these problems were rectified the system was still able to achieve performance goals only when refreshed regenerate brine was used.

The system design had anticipated calcium carbonate scaling within the ARRP system. In an attempt to limit this occurrence soda ash was not used for non-carbonate hardness removal in the brine clarification process. Additionally, the Reuse Plant regenerate basins were enclosed to preclude the exchange of carbon dioxide from the air. It was presumed that the scaling would be carbonate limited and calcium level was allowed to climb. This approach was partly successful. Scaling did not occur within the ARRP system throughout the startup.

Prior to initiation of the preliminary process evaluation phases of plant operation ion exchange performance deteriorated. Investigation of the *clinoptilolite* media revealed that the scaling which had been avoided in the ARRP system was instead occurring within the media itself. The *clinoptilolite* media had accumulated approximately 20 pounds per cubic foot of calcium carbonate. Ion exchange was necessarily deleted from the first process evaluation phase (Phase 1) while remediation steps proceeded.

Scaling within the media had occurred as a result of the high calcium pH brine solution contacting rinse water which contained excessive levels of bicarbonate ions. Before operation of the system could be resumed the media needed to be cleaned. Mild acid cleaning was successfully conducted with minimal loss of material. To prevent a reoccurrence of the scaling, calcium and pH control were instituted in the brine regeneration circuit. Calcium was controlled by the addition of soda ash in the brine clarification step.

The precipitation begun in the clarifier unfortunately continued throughout the brine circuit. Severe scaling plagued the system operations thereafter appearing in: the clarifier itself, clarifier effluent piping, pumps, control valves and the media in the desorption tower. Following desorption of the ammonia, the recovered brine pH was adjusted with concentrated hydrochloric acid using a retrofit of a peristaltic tubing pump. Scale formation was inhibited by maintaining recovered brine pH between 8.5 and 9.0.

During startup operations it was determined the system was unable to achieve the design goal of less than one part per million ammonia-nitrogen. One reason was that the influent ammonia concentration fluctuated over an unanticipated wide range (40% of the time in excess of the 22 ppm). In addition, the brine regeneration processing rate was found to be the factor limiting the overall cycle duration. The minimum regeneration cycle was established at 12 hours which dictated a service cycle of 24 hours. This increase in total cycle time by 50% mandated a fixed cycle time approach to system operation, and effluent ammonia concentrations were allowed to float for the remainder of the ion exchange operation.

The Selective Ion Exchange process was operated and evaluated during phases 2 and 3 of the testing period. The results of this extensive testing revealed that the process was highly complex (ARRP requires 40 steps, 5 chemicals were used, more than two dozen pumps, and several large motors) and subject to malfunctions. The cost was also high, both in capital and operation, for the removal of a single contaminant with no health significance. Due to these findings the ion exchange process became a candidate for replacement.

Alternative ammonia removal processes were considered included breakpoint chlorination, air stripping, biological nitrification and denitrification, reverse osmosis, and blending. Breakpoint chlorination has the advantages that it completely removes nitrogen and is independent of temperature. It's not expensive but free chlorine use was a concern since this may create chlorinated organic byproducts some of which were regulated. Also, the expected high residual chlorine concentrations this process would require dechlorination.

Air stripping was temperature dependent and there were operational complications caused by scaling at high pH. Also, there were concerns that ammonia gas discharge might be classified as an air pollutant.

Biological nitrification and denitrification was a proven process for wastewater treatment plants that require ammonia removal. The capital costs were relatively low and all the nitrogen can be removed using this process. Another advantage was that it appeared that the wastewater treatment facility supplying the secondary treated wastewater for the Reuse Plant would be required to remove ammonia using this process in the future.

Reverse osmosis was a process that can remove ammonia and it was already included in the treatment plant. Adding another reverse osmosis treatment step for ammonia removal was not practical.

Biological nitrogen removal was, consequently, selected for more study since it appeared to be the best alternative process. Pilot projects were undertaken as research studies conducted as part of the requirements for a Master's degree by the Plant Supervisor at the Environmental Engineering Department at University of Colorado. These verified that the biological nitrogen ammonia removal process was both reliable and cost effective.

Because of the process and operational concerns, a decision was made that selective ion exchange should not be used in a future full-scale reuse facility. Additionally, the demonstration Project should proceed relying on reverse osmosis (already part of the treatment process) for ammonia removal. Biological nitrification and denitrification was expected for a proposed full-scale treatment facility.

The Demonstration Project goals were not compromised by proceeding without adding biological nitrification and denitrification since the animal health effects studies focused exclusively on organic substances and ammonia has no known long-term health consequences.

Also, it was highly likely that any future full-scale reuse facility would obtain its influent water from a wastewater treatment facility that was nitrifying and denitrifying its effluent in any case.[2]

## Activated Carbon Adsorption (1st stage and 2nd stage)

Two stages of granular activated carbon (GAC) adsorption were provided in the Reuse Demonstration Plant. The first-stage was sized to treat the entire 1 mgd flow while the second-stage was sized to treat the reduced flow stream of 0.1 mgd. The goal of the first-stage activated carbon process was to remove the dissolved organics present to a level not to exceed a TOC of 4 mg/L. Carbon contactors were operated in down-flow mode with the flow and pressure control downstream. The carbon used during all phases of operation was Filtrasorb™ 300 (Calgon Carbon Corp., PA). This carbon was a crushed coal-based media with an 8 x 30 mesh size range.

Phase 1 used virgin carbon in the first age carbon contactors and Phase 2 operation used regenerated carbon. The second-stage carbon adsorption process initially followed ozonation sandwiched between the two carbon treatment stages and was used primarily as a polishing process for dissolved organic carbon.

The Phase 3 organic removal sequence omitted first-stage activated carbon but included ozonation followed by one stage (second-stage) of activated carbon adsorption. The ozone treatment was used to both oxidize organics and to provide oxygen used to operate the activated carbon treatment that followed as a biological contactor.

During a Phase 4 mini-study reverse osmosis and air stripping were used as the only treatment processes for organics removal. This operation only lasted two weeks before activated carbon adsorption was inserted upstream of reverse osmosis. This was required because of rapid pressure increase experienced in the reverse osmosis system.

The activated carbon systems, both first- and second-stage, exhibited changing TOC removal percentages with new or regenerated carbon. The TOC was removed by about 90% initially and then tapered to about 45% removal (Figure 6-1). It was observed that in both first-

---

[2] Authors note. At the time of publishing this report the wastewater treatment facility has already reduced the effluent ammonia concentration by about half since the Demonstration Project was completed and is anticipating lowering this value by as much as a factor of ten in the future.

stage and second-stage activated carbon after about twenty weeks of operation the TOC removal had reached the steady state removal of about 45% to 50%.

**Figure 6-1**
**Activated Carbon TOC Removal Performance**
**Column Startup**

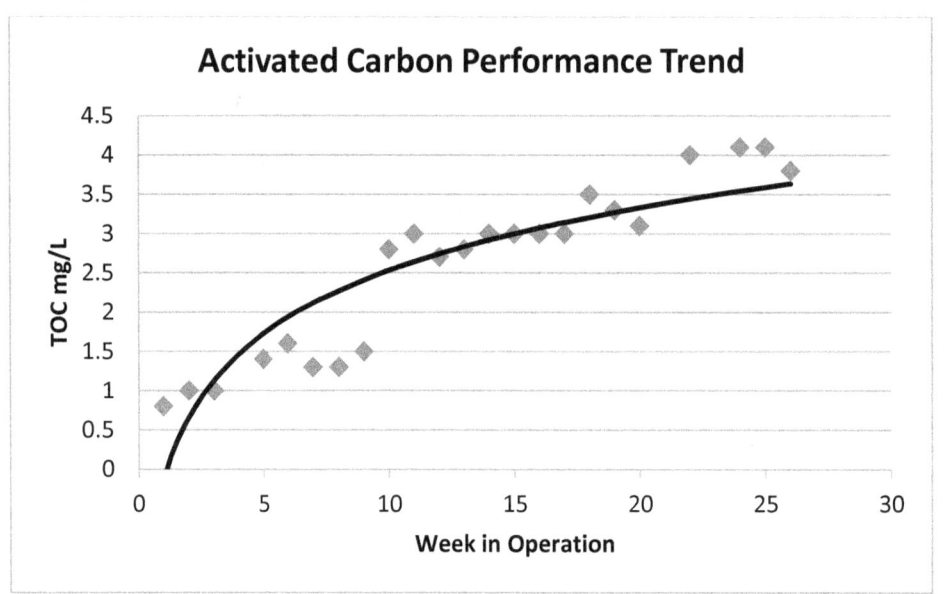

During phase 3 only one stage of activated carbon was operated and it was positioned downstream from ozone treatment (second-stage carbon position). Ozone did not appear to enhance the operation of the GAC column. This may be explained because the second-stage carbon had already been in operation and was already inoculated with bacteria.

The GAC treatment during phase 3 demonstrated the removal of a broad spectrum organic compounds including many volatile compounds. There were only eleven identified volatile compounds that were present in sufficient concentrations to evaluate removal effectiveness. All the eleven VOCs were removed by the single-stage carbon treatment to levels below 1 μg/L except for chloroform and tetrachloroethylene. The absorption/desorption cycle observed earlier during phase 2 of operation for these compounds was experienced again in phase 3 (Figure 7-1).

During one of the mini-studies that comprised Phase 4 the plant treatment sequence omitted reverse osmosis. In this configuration chlorine dioxide disinfection followed activated carbon adsorption. After about two weeks of operation the chlorine dioxide demand in the plant effluent increased dramatically. Investigations verified that nitrite coming from the activated carbon column was responsible for the increased demand. Ammonia residual from the ion exchange process was stimulating nitrification within the carbon columns but limited oxygen levels resulted in the production of nitrite rather than nitrate. The flow was reduced through the ion exchange thus reducing the ammonia residual and the nitrite concentration decreased immediately. By manipulating the ammonia and oxygen levels of the carbon system influent, it was possible to control the nitrite production from biological activity in the carbon columns. Once a decision was made to use ozone as the primary disinfectant and to move it after reverse osmosis treatment, there was no need to use this operational strategy, since chlorine dioxide was no longer being used.

As explained earlier, during phase 4 several treatment sequences were evaluated. The first was to rely solely on reverse osmosis and air stripping for organic removal. Immediately it was clear that the reverse osmosis system could not operate in this configuration due to excessive pressure drop that occurred across system and extremely high operating pressure. Unacceptably short operational periods between cleanings required action. Additional pretreatment was needed ahead of RO to control this situation.

A single-stage of activated carbon was added before the reverse osmosis. This reduced the operating pressure in the RO process to an acceptable level. This modified process sequence was run for about 81 days before the RO system again experienced operational difficulties due to an increased pressure drop. At this time the ozone process was put back into operation upstream of the carbon column. This did not alleviate the pressure drop issue but the RO operating pressure was reduced so the process could be used to produce high quality water by enduring frequent cleanings.

During the process assessment phases free-living microscopic nematodes were detected in the carbon process effluent. Nematodes were also found in the plant influent and after ozone treatment. One explanation for this was that the cysts apparently resisted lime treatment and chlorine dioxide disinfection before filtration. The organisms then penetrated filters, ion exchange media, and survived ozonation, and penetrated activated carbon filters. This series of events was highly unlikely and instead it is probable that these organisms were transferred to the various locations by unintentionally using untreated water throughout the plant during configuration changes and non-sanitary operational procedures. The viability of the organisms was not determined. Nonetheless, alive or dead the nematodes could shield pathogens from final disinfection. For this reason, and since it was not clear how they move through the processes, it became prudent to incorporate a membrane filtration step to block these organisms from the plant effluent if they were present.

After assessing the results from the treatment process evaluations, the primary organic removal processes selected for use during the whole-animal health effects testing program were a single-stage of granular activated carbon adsorption followed by RO, and air stripping. The carbon columns would be operated at a 0.1 mgd flow rate which resulted in an empty bed contact time of 43 minutes. Carbon columns were backwashed when the head loss reaches 15 feet.

Just before reconfiguring the plant into the final health effects treatment sequence ultraviolet irradiation was tested upstream of the single-stage activated carbon adsorption treatment. This evaluation was to see if the biological growth experienced in the activated carbon column could be controlled so that the frequency of backwashing, due to the high rate of head loss, could be reduced. The UV operation, to be discussed later, was effective in removing biological organisms but the impact on the activated carbon operation was not fully understood before beginning the health effects testing program. Later, as the health effects study treatment continued, this treatment step was found to be ineffective in extending the time between carbon column backwashing. It is now apparent that UV would not be necessary in a full-scale treatment plant.

## Ozonation

The ozone process was sized to treat 0.1 mgd flow rate. This flow was initially adjusted by the level control valve located in the discharge piping of the ion exchange columns. The remaining 90% of the plant influent was not treated with ozone but was disinfected with chlorine dioxide and collected in a basin for industrial and landscape irrigation uses.

The ozone contact basin was rectangular with a depth of 15.8 feet the length of 14 feet and the width of 2.5 feet. It was divided into six compartments followed by a quiescent zone. Aluminum baffle plates directed the water to flow in a vertically serpentine (over and under) pattern. At the bottom of the first six compartments diffusers bubbled the ozone gas up through the incoming water. The quiescent zone located at the end of the basin had no diffuser. The primary purpose of this zone was to allow the release of ozone bubbles that have not been absorbed in the water. The entire basin was under a slight negative pressure so that gases released could be collected and sent to a catalytic ozone destruction unit (this converted ozone to oxygen).

The Reuse Plant ozone process used ambient air as a source of oxygen for ozone production. Air was first compressed to a pressure between 100 and 103 psig. It then passed through filters to remove oily substances. Next a desiccant was utilized to remove moisture from the air giving a dew point of approximately -70°F. Finally the air passed through another set of filters to remove dust and any desiccant particles before entering the ozone generator at a pressure of about 11 psig.

Ozone was applied to the contact basin from one of the two generators. The mode of operation was such that one generator was operating while the other generator was in standby. Both ozone generators were constructed of stainless steel cylindrical tanks. Within each tank 15 glass dielectric tubes were encircled by an equal number of stainless steel tubes. A discharge field generated a corona effect in the gap between the glass and the steel tubes. As air was pumped through, the oxygen molecules were split and ozone was formed. The production of ozone was varied by the voltage regulator on each generator that controls the amount of electricity to create the corona discharge.

The main purpose for ozone being applied to the flow ahead of activated carbon treatment was to enhance organic compound removal. Secondarily, ozone was included in the treatment sequence as an additional disinfectant.

Since the effect on activated carbon performance by ozone was negligible, and due to the need for a powerful disinfectant in the treatment sequence, ozone was retained as an important step in the treatment process for the animal health effects study. It was moved though from a point ahead of second-stage activated carbon to follow the reverse osmosis and air stripping processes. In this position it would serve as a primary disinfectant for the reuse treatment sequence. The combination of ozone as primary disinfectant and monochloramine as a residual disinfectant satisfied the requirements for drinking water disinfection. Ozone was a well-known disinfectant used in water treatment and was suitable for this purpose. Ozone application following reverse osmosis required only a minimum residual of 0.5 mg/L to accomplish effective disinfection.

## Reverse Osmosis

The design of the Reuse Demonstration Plant employed a multiple barrier concept where no one process was relied upon for complete removal of any contaminant. During conceptual design the reverse osmosis system was inserted, in part, into the treatment process as an ammonia polishing process. Pilot studies had demonstrated that with proper selection of membranes ammonia removals through the reverse osmosis system could exceed 90%. Additionally, it was known that a membrane process would provide a final physical barrier to other contaminants including dissolved salts, bacteria, viruses, and organic compounds. Project objectives such as achieving a total dissolved solids content equal or below that of Denver drinking water would be difficult to meet without a desalination system such as reverse osmosis.

The reverse osmosis system was designed to process 0.1 mgd at 90% recovery. The feed flow was split evenly between two parallel units. A third unit was out of service for cleaning or in standby. Each unit was configured in a 4-2-1 array. In this configuration the reverse osmosis feed water was divided evenly among four vessels that make up the first pass. The product water was collected in a manifold system and combined with the product water from the successive passes. The brine waste from the first pass was divided evenly as feed water among two vessels that make up the second pass. The brine from the second pass was the feed water for the single vessel that makes up the third pass. Fluid Systems® model 4600 spiral wound polyamide membranes were used. No other elements were evaluated. Feed water was supplied through a 21 stage centrifugal pump that produced a pressure of approximately 550 psig. The product water was transported to the air stripping tower while the brine stream was returned to the wastewater treatment plant for processing.

Cartridge filters of 5 μm were used to filter carbon fines from the reverse osmosis feed water. The feed water pH was also adjusted to 5.8 with hydrochloric acid while sodium hexametaphosphate was applied as a scale inhibitor at 5 mg/L.

The process was monitored continuously for pH, turbidity, temperature, conductivity, pressure, and flow. Safety interlocks shut the system down when pre-established limits were exceeded. The reverse osmosis system operated within guidelines specified by the system manufacturer but often at less than design flows and recoveries. Operation at lower flows resulted from periodic fouling problems that were difficult to resolve.

Cleaning of the system was initiated with feedwater when the pressure increased by approximately 10%. The cleaning procedure used a citric acid solution that had been adjusted to a pH of 3.5 followed by a solution of borax, trisodium phosphate, and EDTA adjusted to a pH 11.0. Each pass of the unit was cleaned individually for one hour with each solution. Alternate cleaning procedures and chemicals were used with varying results. Out of service units were stored in a solution of 0.25% formaldehyde.

The feedwater pH of the reverse osmosis system was adjusted to 5.8 with hydrochloric acid. The original system design included sulfuric acid for pH adjustment. Sulfuric acid use increased sulfate concentration in the feedwater and the potential for irreversible calcium sulfate fouling. The acid feed system, thus, was modified to use hydrochloric acid for pH adjustment. This switch involved changing the components of the feed pumps to make them chemically compatible with the hydrochloric acid.

Storing the reverse osmosis system components while they were out of service in a formaldehyde solution required some system modifications. The procedure, used after startup, was initially ineffective because most of the formaldehyde solution would run out of the unit and down the drain within a few hours. After the system was modified to incorporate isolate valves and a head tank this procedure was effective.

Reverse osmosis was selected to remain in the reuse plan treatment train for the animal health effects study because of its multi-contaminant removal capability. It provided a final barrier to dissolved and particulate matter, bacteria, virus, and a large range of organic compounds. The ammonia removal of the reverse osmosis system provided a polishing step for upsets and biological ammonia removal processes that may be upstream in a future full-scale treatment plant. The reverse osmosis system provided an additional contaminant removal barrier safety factor that few other systems can match.

The experience gained in operating the reverse osmosis system led to the implementation of several improvements and recommendations for a full-scale plant design that include:

- RO systems should use isolate valves on the product lines and have a storage head tank for a disinfection solution.
- Effective disinfectants that are compatible with materials of construction need to be developed for use in full-scale reverse osmosis systems.

- A surge suppressing device should be installed on the discharge of the high-pressure feed pumps to suppress the hydraulic anomalies associated with starting the pumps.

## Air Stripping

The air stripping process served a dual purpose in the reuse treatment train. Air stripping not only removed dissolved carbon dioxide from the water following reverse osmosis but it removed volatile organic compounds as well. The air stripping system was located directly downstream of the reverse osmosis system. The air was fed undercurrent to the water stream at a gas to liquid volume ratio of 100 to 1.

The air stripping tower's primary function was decarbonation. Although carbon dioxide was not measured directly, the effect of air stripping shifted pH across the tower. Typically the pH of the reverse osmosis product water was less than 5.0. The pH of the air stripper water was about 6.6. The dissolved oxygen also increased about 1.0 mg/L after air stripping (these values fluctuated with water temperature).

The removal of volatile organics by air stripping was desirable because the reverse osmosis membranes were somewhat transparent to these compounds. For that reason, air stripping was recommended as one of the treatment processes to be included in the animal health effects study treatment sequence.

## Residual Disinfection (chlorination, chlorine dioxide, or chloramination)

The preliminary design of the reuse demonstration facility treatment sequence included breakpoint chlorination bracketed by two stages of activated carbon adsorption. In the conceptual design, chlorination was to function both as a disinfectant and as an additional removal process for residual ammonia from the upstream selective ion exchange process. The conceptual design also included ozonation as a disinfectant and oxidant for refractory organics following membrane separation and preceding the final disinfection with free chlorine.

As the plant design evolved, breakpoint chlorination was removed from the treatment sequence because of concern over the formation of trihalomethane's and other chlorinated compounds. Concurrently with this change, ozonation was relocated upstream to occupy the position previously held by breakpoint chlorination. In this manner ozone was retained as a pathogen barrier. The impact on refractory organics was assessed by the performance of the second-stage activated carbon. Also, at this time, chlorine was eliminated as a final disinfectant. Again the concern was for the formation of toxic byproducts. Chlorine dioxide was substituted as the final disinfectant since it did not form chlorinated organic byproducts while it provided an effective residual for distribution system requirements.

While not discussed in the plant design, the experimental plan for the reuse facility included the evaluation of residual disinfectants that would be compatible with Denver's existing distribution system. Since the Denver system relied on chloramination, it was anticipated that this chemical would be studied as the possible residual disinfectant.

### Chlorine Dioxide

Chlorine dioxide was generated as needed using the reaction of sodium chlorite with chlorine gas. Several generators were performance tested at the reuse facility. The last generator used at the Demonstration Plant incorporated an improved design. Vacuum eduction of the reactants made this system much faster in forming chlorine dioxide and thus, reduced the formation of undesirable by-products.

Chlorine dioxide demand varied by as much as an order of magnitude as a consequence of nitrite production from the second-stage carbon columns. A further complication of nitrite occurrence was its interference with the colorimetric test used for measuring chlorine dioxide residual. This finding explained some positive coliform results that occurred even though a seemingly adequate disinfectant residual was present. After this discovery, the operational setpoint for chlorine dioxide feed was raised to 0.25 mg/L to ensure adequate disinfection when nitrite was present. The average dosage in the plant activated carbon product water was 2.2 mg/L and the average residual for the RO product was 0.20 mg/L and contact time averaged 14 minutes.

Chlorine dioxide was used as an intermediate disinfectant upstream of filters during some phases of the plant operational testing. The purpose was to reduce microbial populations on a continuous basis as opposed to the batch disinfection methods which had been practiced previously. Chlorine dioxide was fed in the influent to the filter at a dosage of about 1.5 mg/L during this period.

The filter pre-disinfection with chlorine dioxide was effective in eliminating microbial loading on the filter media but turbidity increased dramatically. Pre-disinfection was discontinued once it was determined that downstream ozonation was reactivating the chlorite residual and recreating chlorine dioxide.

Chlorine dioxide demonstrated its effectiveness as a disinfectant for water reuse. Following several improvements, the generator was able to be adequately controlled and monitored. The final product water microbiology compared favorably with Denver drinking water. But, given the unresolved questions about the toxicity of chlorine dioxide and its reaction byproducts (chlorite and chlorate), chlorine dioxide disinfection was deleted from the final plant health effects treatment configuration. Chlorine dioxide should, nonetheless, be included in a future full-scale reuse treatment plant so that it can be used for disinfection of off-line processes like filtration.

### Chloramine

Free chlorine was available within the Reuse Plant since it was used to generate chlorine dioxide. So, it could be used to combine with the ammonia residual in the reuse product water to form monochloramine. This process was used in the final health effects treatment sequence to provide a residual disinfectant. There were several reasons for this including compatibility with Denver drinking water which had used chloramines in the distribution system as a secondary residual disinfectant for more than 70 years. Also, chloramines did not form disinfection byproducts that were of concern with free chlorine.

## Ultraviolet Irradiation

Prior to the reconfiguration of the plant into the health effects treatment sequence, ultraviolet irradiation (UV) was tested as an intermediate pathogen barrier. The first installation was located upstream of reverse osmosis and downstream of activated carbon adsorption. The UV contactor construction consisted of a spare UV element inserted into a straight run of three inch steel pipe. Initial performance as measured by coliform reduction was encouraging and subsequently, a manufactured package disinfection system was purchased. This unit was installed upstream of activated carbon adsorption to reduce the bacterial populations which had caused head loss within the carbon media. While performance of UV as measured by microbial reductions in the water was impressive, its impact on activated carbon head loss was not conclusive. In spite of this, UV was operated during the health effects treatment upstream of activated carbon to provide an added pathogen barrier.

Upon reflection following completion of the two-year animal health effects study, it was clear that ultraviolet disinfection upstream of activated carbon was not necessary to provide potable water quality from secondary unchlorinated sewage effluent. Its use did not improve the performance of activated carbon. Ultraviolet irradiation was not recommended for a future full-scale potable reuse treatment plant.

## Ultrafiltration

Ultrafiltration (UF) was evaluated as a possible parallel treatment process to reverse osmosis. Ultrafiltration provided a membrane barrier desirable for the potable treatment system. While it did not remove dissolved salts and, thus, would produce water higher in these non-regulated substances than Denver drinking water, it would provide product water suitable for blending with the reverse osmosis treatment stream. A 50/50 RO and UF blend would result in more stable water that was almost identical in quality to Denver's current drinking water supply.

A Desalination Systems Inc. ® G10 membrane (molecular weight cutoff 2500) was selected for evaluation after testing several competitive membranes. The material was polysulfone with a fiberglass wrap. Effluent from activated carbon adsorption was used as the feedwater for initial evaluations and later it was placed in the same position in the treatment sequence as reverse osmosis. The ultrafiltration system was fed using a 1/4 hp pump through a flow control valve, a pressure regulator, flow meter, and then the membrane. Effluent from the system was divided between the permeate and waste streams. The recovery of the ultrafiltration system was maintained at 85% or more to model the nominal reverse osmosis conditions.

The ultrafiltration performance tests showed that turbidity was reduced to the same level as reverse osmosis. The UF system also reduced total organic carbon and microbiological organisms. The total organic carbon was reduced by over 50% and microbiological organisms were completely eliminated. The cost of operating the UF system was found to be one fourth that of the reverse osmosis. The UF system proved to be an effective alternative or a parallel system to reverse osmosis in the Reuse Demonstration Plant, and it was included on a pilot-scale in the final health effects testing sequence.

## Carbon Regeneration (fluidized bed furnace)

The design of the Reuse Demonstration Plant included the thermal regeneration of activated carbon using a fluidized bed furnace. The regeneration system was comprised not only the fluidized bed furnace but also two large storage tanks, slurry transport piping, pumping, and controls for the transfer of carbon to and from the contactors.

The furnace system consisted of five main components: carbon transport, drying, regeneration, incineration, and off-gas treatment. Carbon transfer included the slurry transport of the spent carbon from the storage tank to the dewatering screw hopper and the eduction of regenerated carbon from the quench tank to the column. Control was by mechanical timers and level probes. The dryer blower fluidized the spent carbon above the dryer grid. Off-gas from the incinerator section provided the drying temperature and fluidizing velocity via a recycle blower. Dryer temperature was controlled at 150° C by a water spray with emergency spray backup. Carbon fines were separated from the recycle flow stream by a cyclone. Feed to the dryer was via a manually controlled variable speed dewatering screw auger. The regenerator section supplied absorbed organics in an oxygen-poor environment. Feed to the generator was metered by a rotary valve which also acted as an air lock between the dryer and the regenerator. Temperature and free oxygen controlled the flow of natural gas. Following gasification the carbon was quenched before transport.

The incinerator functioned to combust the off-gassed organics at a temperature of 900°C in the presence of 3 to 4% oxygen. A part of this atmosphere was withdrawn or recycled through the dryer bed and the remainder was exhausted. Temperature and free oxygen regulated natural gas and airflow at the second burner. The off-gas treatment consisted of scrubbers to cool the exhaust and remove the particulates. The exhaust was pre-cooled before entering a variable throat venturi scrubber. The gases then passed through a tray impingement scrubber before the exhaust blower exhausted through the roof.

The furnace system construction was not completed until after the rest of the plant was functional. Even after completing many modifications, the furnace still did not operate in a completely automatic mode. This caused a safety concern since operation could only be accomplished in semi-manual control. Several batches of used carbon, nonetheless, were regenerated using the fluidized bed furnace after it had been retrofitted and tested. Although the carbon regenerated satisfied the quality goals for this process, the complexity and cost of this system made it a candidate for elimination from a future full-scale facility.

The need for an on-site carbon regeneration furnace was also questioned because of the finding that the activated carbon columns could be operated for lengthy periods without regeneration. Activated carbon performance during the operational testing phases, and continuing during the animal health effects study, exhibited a 50% steady-state organic removal for more than two years (Figure 6-2). This finding supported the recommendation to perform infrequent carbon regeneration by a specialty contractor off-site.

**Figure 6-2**
**Activated Carbon**
**% TOC Removal**

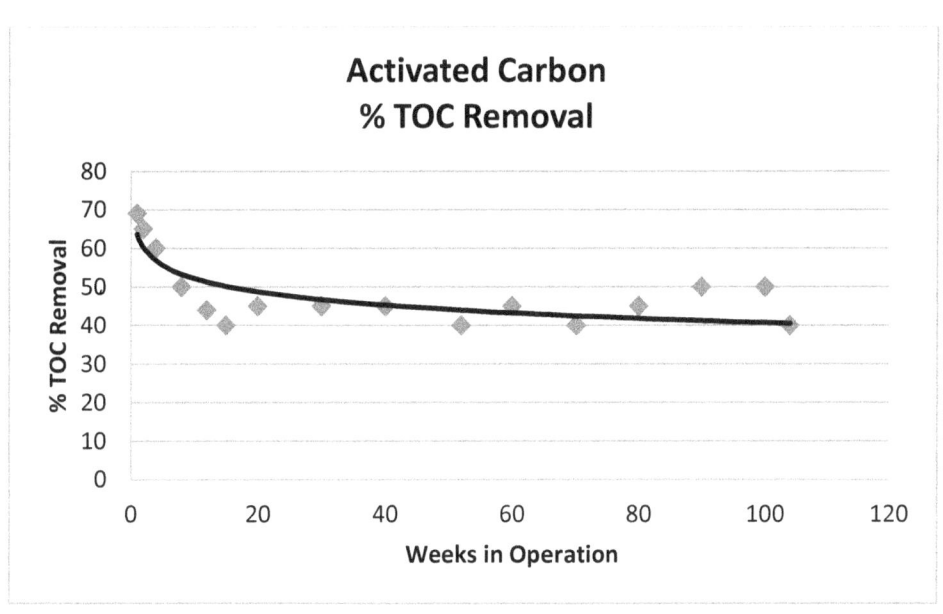

# 7

# Special Studies Summary

Several special studies were conducted during the process assessment phases and even during the health effects treatment operational period. These investigations spanned a broad spectrum of topics that addressed problems encountered in plant operation or to evaluate options supporting possible plant performance improvements and laboratory studies. The summaries in this section briefly describe the results of these investigations.

## Reverse Osmosis Process Related Studies

### I. Optimizing Activated Carbon Column Backwash

Second-stage activated carbon columns at the Reuse Demonstration Plant were positioned upstream of the reverse osmosis process. This study investigated backwash procedures to reduce the impact on reverse osmosis.

Down-flow rinsing, blending, and resting were three operational procedures investigated to reduce the impact of the backwash of activated carbon on reverse osmosis units that followed. Resting was not found to be beneficial. A combination of down-flow rinsing and blending produced the best results

The optimum procedure included a 30 minute down-flow rinse of the standby carbon column followed by a 30 minute operation of the standby column blended with the online carbon columns before they were removed from service. The standby column was then put into service while concurrently switching the online column to standby status. Backwash of the exhausted column was then initiated followed by a down-flow rinse.

In a full-scale treatment plant a backwash supply wet well should be provided (with appropriate supply pumps) to eliminate issues created by backwashing with process water.

### II. Reverse Osmosis Cartridge Pre-filter Comparison Testing

Cartridge filter elements (5 µm) were used as a pre-filter on the reverse osmosis system to remove the carbon fines coming from the activated carbon process. The cartridge filters were located upstream of the RO feed pumps. Any pressure drop across the cartridge filters then reduced the net suction head on the feed pumps.

The Filterite® filters used during the plant startup provided a consistent duty cycle. They performed as well as the HydraX ® filters evaluated as an alternative but operated at a much lower head loss. The lower head loss experienced with the Filterite® filters resulted in a longer service cycle and offset the cost reduction of using the less expensive HydraX® filters.

### III. RO System Disinfection

During the process assessment phase of the Demonstration Project bacteria were detected in the reverse osmosis effluent. An investigation subsequently commenced to determine if coliform bacteria had colonized the permeate side of the RO membranes. Because the membrane surfaces could not be tested directly, the permeate piping was tested. These tests verified the presence of bacteria.

A procedure provided by the manufacturer was initiated to disinfect the entire RO system. The units were taken out of service, cleaned using the normal cleaning procedures and the elements then filled with formaldehyde using the cleaning system components. The formaldehyde solution (0.25% by weight) was recirculated for 15 minutes every two hours for approximately 3 days. This increased the membrane exposure to the formaldehyde.

The membranes were also removed from the pressure vessels so that the inside of the permeate tubes and the fiberglass casings could be cleaned and disinfected. After cleaning, the elements were reinstalled in the pressure vessels and the system returned to service. Bacteria were still found, particularly in the third pass of the reverse osmosis pressure vessels. This was anticipated since this system did not contain isolate valves on the permeate lines. This allowed the formaldehyde to run out of the product drain line. Membrane contact with the formaldehyde was minimal.

After this discovery, isolate valves were installed along with a formaldehyde head tank to ensure the entire system was full of the solution. The product drain valve was closed, thus holding the solution in the vessels. After correcting this deficiency and repeating the disinfection procedure, analysis showed no coliform or background colonies. The formaldehyde residual after 24 hours remained at 0.25% for the entire period. This modification eliminated the bacterial colonization problem in the RO system. The results highlighted the importance of developing effective disinfectants that are compatible with membrane materials so they can be used routinely.

### *IV. Reverse Osmosis fouling and cleaning developments*

After more than a year in operation the service cycles between RO cleanings declined from four weeks to less than a week under constant flow and recovery operating conditions. This operational problem occurred in all three of the reverse osmosis arrays indicating a system-wide problem. Organic substances trapped on the membranes apparently caused this unacceptable performance. This preliminary finding was corroborated with observations from an element autopsy report.

In an attempt to improve performance the cleaning regiment was modified several times. Reversing the order of the cleaning solution application and using enzyme cleaners proved unsuccessful. Air scouring (termed *scrubbling*) was helpful as a method of dislodging the organic slime trapped within the elements.

Plugging factor tests (SDI) provided little information about fouling potential of the feedwater. The performance changes experienced within the RO systems were not predicted by the plugging factor tests; so plugging factor testing was suspended.

There were many unexplained changes in operating performance of the RO system. These changes did not correlate with plant operation or feedwater quality. Although membrane fouling eluded a permanent solution, the elements within the RO system remained in continuous use for almost four years and overall performance of the system was acceptable. Fouling problems led to increased operating costs. This issue received continued investigation during the health effects operational phase.

### *V. Feasibility of Operating the Reverse Osmosis System at 95% Recovery*

The research conducted at Denver's direct potable water Reuse Demonstration Plant has shown that a future full-scale potable water reclamation plant should include a reverse osmosis system. It was also clear that a reverse osmosis system would require a larger capital investment and have higher operating costs than any other treatment process.

Much thought was given to ways to reduce these costs. The economy of scale and a split treatment process sequence utilizing ultrafiltration membranes substituted for reverse osmosis would reduce construction and operating costs for full-scale plant implementation. New generation lower-pressure reverse osmosis membranes and energy recovery systems were also expected to lower operational costs even more. Each of these cost-saving ideas was likely to be implemented in a full-scale facility.

A method of reducing both capital and operating costs that had not been previously considered involved increasing the recovery of the reverse osmosis system from 90% to 95%. As a result of this modification water production would increase by 5% while the volume of the waste stream would be decreased by 50%. The size of both the reverse osmosis system and the brine disposal equipment could then be reduced. The consequences of increasing the recovery include a possible degradation in product water quality, a higher potential for chemical fouling, and higher operating pressures that translated to higher pumping costs. The extent to which the benefits outweigh the liabilities are best examined by a process performance computer model in conjunction with data collected from a system operating at 95% recovery. This combination was used in evaluating the feasibility of operating a 95% recovery system.

In reviewing the data, from both computer analysis and the operating plant 95% recovery pilot-size unit, it was concluded that it was feasible to operate a reverse osmosis system for the reuse application at 95% recovery. A 95% recovery system will lower both capital and operating costs in a full-scale facility without a significant degradation in water quality. The primary factors evaluated by the computer analysis and the 95% operating pilot system included water quality, fouling potential and actual fouling, operating pressure, and cost. From this evaluation, it was recommended to include a system capable of operating at 95% recovery for the future full-scale plant.

# Disinfection Process Investigations

## I. Chlorine Dioxide Demand Investigation

Chlorine dioxide was employed as a primary disinfectant for the Reuse Demonstration Plant. During one treatment sequence configuration activated carbon was upstream of chlorine dioxide. Reverse osmosis was omitted from this process sequence. After only 10 days of operation in this configuration chlorine dioxide demand began increasing. The rate of increase was such that the demand was not satisfied on several occasions which resulted in some positive coliform test results.

Biological nitrification was occurring in the activated carbon columns and this produced elevated concentrations of nitrite. This created an increased chlorine dioxide demand. A 10:1 weight ratio of chlorine dioxide to nitrite was necessary to oxidize the nitrite that was present. Monitoring nitrite and then adding enough chlorine dioxide to satisfy the 10:1 weight ratio requirement resolved this problem. Further, the nitrite level coming from the carbon columns was reduced or eliminated by manipulating the oxygen level coming from the filters ahead of the carbon system.

## *II. Chlorine Dioxide Reaction By-Products Removal by Activated Carbon*

Chlorine dioxide, upon reaction with substances in the water, can produce chlorite and chlorate as byproducts. Both of these byproducts are regulated in drinking water. It was necessary then to remove these byproducts if they occurred in concentrations nearing regulatory limits. Also, removal of these byproducts would allow higher chlorine dioxide dosages without concern for exceeding regulatory limits.

Activated carbon can remove chlorite and this treatment process was used in the Reuse Plant so it was evaluated for this purpose. One concern was that carbon has a limited capacity for chlorite reduction. Based on bench scale experiments, at chlorite concentrations of 4 mg/L, a six-month plant-scale exposure would result in approximately 20% loss of GAC capacity.

Actual plant average dose of 1.5 mg/L chlorine dioxide should allow uninterrupted operation for up to a year before significant loss of GAC capacity would require regeneration. Bench scale experimental results also showed a need for adequate reaction time with carbon for optimal chlorite reduction.

The positive aspects of chlorine dioxide as a pre-disinfectant found in this study include increased filter runs and the addition of another barrier for microorganisms. The drawbacks were increased turbidity of the filter effluent and the deleterious effect on ozone off gas treatment. (Chlorine dioxide was recreated by the reaction of ozone and chlorite. This chemical then attacked the catalyst used in the ozone destruction process.)

These findings combined with the concern from chlorine dioxide by-products lead to a decision to omit it from the Reuse Treatment Plant process sequence. The many benefits of chlorine dioxide as an effective disinfectant, however, supported its inclusion in a future full-scale reuse treatment plant to use off-line when necessary.

### III. Optimization of the Ozone Process as a Microbiological Barrier

The ozone treatment process was positioned between first- and second-stage activated carbon adsorption for many of the process sequence evaluations. The purpose of ozone was twofold, to act as a biological barrier and to oxidize organics still present in the flow stream for improved adsorption on the second carbon stage. The main purpose of this optimization project was to develop a correlation between applied ozone dosage and the reduction of microbiological organisms by the ozone process that could then be used to adjust the dosage to meet anticipated demand.

Total organic carbon (TOC) test results, suggested by the literature, were evaluated as a predictor of the optimal ozone dose but this was found to be inaccurate. A correlation was found between ozone dosage, alternatively, and microbiological reduction across the process. As an example, for water containing total coliform of around 400/100 mL an applied ozone dose of 3.4 mg/L was found to be optimal. To effectively apply this method of predicting the ozone dose frequent bacteria testing would be necessary.

### IV. Filter Disinfection with Chlorine Dioxide

Chlorine dioxide was used to shock treat various process components which had become colonized by excessive bacteria populations. One example of this application was the disinfection of the filter media. During the early plant testing it was observed that dissolved oxygen (DO) was being consumed across the filters. This phenomenon was somewhat mystifying since the influent dissolved oxygen concentration was increased to near saturation by an air diffuser system installed in the ballast pond just ahead of filter pumping. After passing through the filters with only minimal contact time the dissolved oxygen decreased to only about 2 mg/L. Activated carbon adsorption followed, and the process operational procedure at the time required the entire column aerobic to provide optimum adsorption characteristics. With influent dissolved oxygen so low this condition could not be assured.

After investigation it was confirmed that excessive bacterial colonization in the filters was responsible for the observed DO reduction. Filters were treated with high dosages of chlorine dioxide 25 to 100 mg/L for durations varying from 2 to 6 hours. A 50 mg/L dosage for two hours was found to be optimum. This treatment was repeated whenever necessary (about every three months).  This procedure not only reduced DO depletion but also kept nitrification within the filters under control.

### V. Chlorine Dioxide Generator Optimization

Chlorine dioxide effectively disinfected all bacterial indicators. The production of chlorine dioxide was highly sensitive to several factors including chlorine dioxide concentration, sodium chlorite concentration, pH, mixing characteristics, and dilution effects. Simple monitoring methods used to determine chlorine dioxide residual were not capable of analyzing the other chlorine species produced in the process. Exacting analytical speciation measurements were needed to accompany yield tests when optimizing chlorine dioxide generators. Without these measurements potentially harmful byproducts, such as chlorite and chlorate, may be unknowingly introduced into the water thus negating the major benefit of chlorine dioxide usage. Thus, the more rigorous amperometric test method was routinely used to monitor chlorine dioxide generator performance.

# Filtration Process Optimization Studies

### I. Polymer Filter Aid Evaluations

Polymers are used in conventional water treatment plants to aid in the removal of suspended particles. At the Reuse Plant, polymers were evaluated to enhance the efficiency of the pressure filters. After testing several polymers, no polymer was found to be effective in lowering the effluent filter turbidity. Certain polymers showed limited degrees of success in lowering turbidity as compared to no polymer these included: Nalco® 8184, and Lloyds® LT24 at dosages of 0.1 ppm and 0.05 ppm respectfully. None of the polymers tested consistently lowered turbidity and none were successful in reaching the filtration goal of 0.1 NTU.

### II. Filter Optimization

Several adjustments were assessed to improve the performance of the filters. These improvements involved optimizing surface wash, backwash rate, and backwash length. The evaluation determined an optimum surface wash time of 17 minutes. A flow rate of 1,250 gallons per minute (gpm) was found to be the optimal backwash rate. And, the optimum length of the backwash was 25 minutes.

Several months after instating these procedures the filter surface was inspected and cracking was noted. This indicated that the backwash settings determined earlier needed to be adjusted. The backwash flow rate was increased to 1,500 gpm and the surface wash duration was adjusted to 20 minutes, backwash period remained at 25 minutes. The optimum filter-to-waste time upon start-up was found to be 10 minutes. These adjustments eliminated the filter surface cracking and improved overall performance.

Turbidity did not break through even at the maximum head loss tested. Still it was determined that filters should be backwashed when head loss reached 11 feet. This conformed to the design criteria and was within the operating ranges of the plant instrumentation and did not present any operational problems.

# Volatile Organic Compound Removal Studies

## I. *Air Stripping of Activated Carbon Effluent*

During phase 3 operation where reverse osmosis was not used, a problem associated with the operation of the activated carbon system was encountered. The breakthrough of low molecular weight organic compounds occurred long before the column was otherwise exhausted. In previous phases of operation reverse osmosis and air stripping followed the carbon process and these processes effectively removed any residual organics.

Air stripping, thus, was evaluated for the removal of remaining volatile organic compounds (VOC). The investigation of the air stripping after activated carbon adsorption and without reverse osmosis only lasted about a month. During that time operational problems including foaming and odors created unacceptable product water. It was concluded that air stripping in this sequence could not be used, thus, the evaluation of its effectiveness in removing volatile organic compounds without prior reverse osmosis treatment was not conducted until a later operational phase.

## II. *Volatile Organic Compound Removal by Activated Carbon*

Granular activated carbon (GAC) effectively removed total organic carbon. The removal efficiency of GAC with respect to specific compounds was affected by fluctuating concentrations of these compounds in the influent water. This investigation examined the removal of a list of volatile organic compounds (VOC) found in the reuse influent water. Activated carbon removed significant amounts of all the VOC's present.

Chloroform was the only compound tested that periodically exceeded 1 μg/L. The so-called chromatographic effect of periodic desorption was observed for chloroform (Figure 7-1) and bromoform. The concentration of chloroform in the GAC effluent exceeded 10 μg/L during one desorption event. Bromoform effluent concentrations remained below 1 μg/L even during desorption. A regular interval of 10 to 11 weeks was determined as the absorption-desorption cycle for these compounds. The chloroform concentration even during desorption was far below any health standard (MCLG = 70 μg/L).

**Figure 7-1**
**Chloroform Desorption from Activated Carbon**

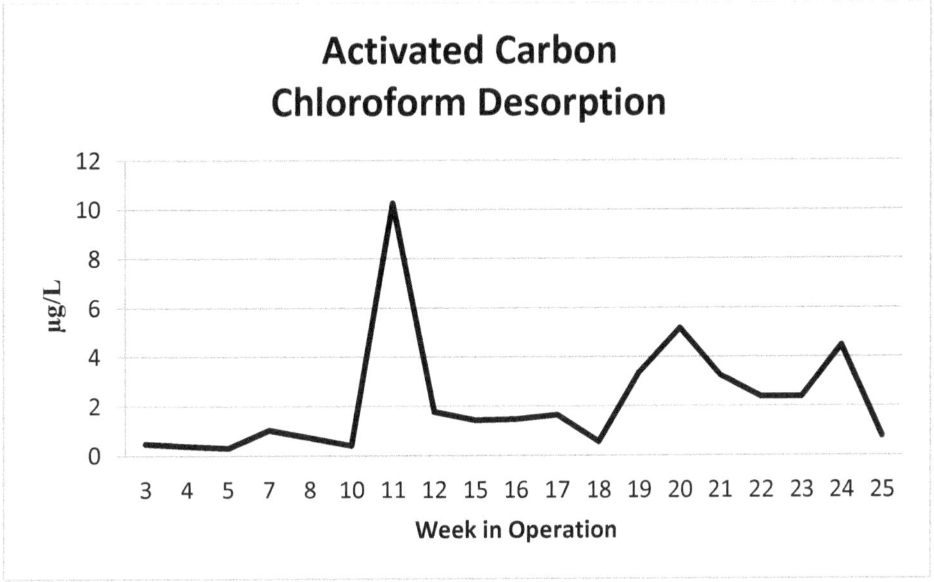

### III. UV/O3- Laboratory Scale Organic Removal Study

Organic removal at the Reuse Demonstration Plant was accomplished by several processes. The combined effect of these treatment steps reduced the total organic carbon levels (TOC) from 17.4 mg/L to less than 0.5 mg/L. A goal of the Project was to produce water which was equal or better than Denver's existing drinking water supply which has a TOC of about 2 mg/L. Several university research studies had demonstrated that photolytic oxidation processes either UV/ozone or UV/peroxide were effective in removing organic substances. A laboratory scale study thus evaluated these processes to reduce organic content and perhaps replace granular activated carbon and reverse osmosis in the reuse treatment system.

Bench testing of the UV/ozone process demonstrated that 200 Watt-minutes of input UV light, and an applied ozone dosage of about 180 mg/L at 90% transfer efficiency, and a pH of 6 would reduce the TOC below the 2 mg/L goal. Using UV/peroxide treatment the goal can be achieved with a UV dosage of 2890 Watt/gallon, a peroxide dosage of 400 mg/L, a solution pH of 2.7, a solution temperature of 50°C, and 27 minutes of treatment.

Both UV/ozone and UV/peroxide treatment would cost $5-$11 per thousand gallons depending on the energy costs and efficiencies of scale. Due primarily to excessive costs, neither UV/ozone nor UV/peroxide treatment processes was competitive with granular activated carbon or reverse osmosis treatment for the removal of organic compounds to reach a TOC of 2 mg/L.

### IV. Halogenated Volatile Organic Compound Removal by Activated Carbon

This study examined the removal of halogenated volatile organic compounds (HVOC) by activated carbon. The second-stage activated carbon was used for this study during a time when it was the only stage of carbon in operation and was the primary process to remove these compounds.

The activated carbon adsorption process was found to be effective removing all the HVOC's present. Effluent concentrations of all compounds were less than 1 µg/L except for chloroform and tetrachloroethylene. The chromatographic effect (the cyclic adsorption and desorption) was observed for chloroform and to a greater extent tetrachloroethylene. Chloroform desorbed during the seventh week of operation.

Apparently caused by the storage and transport of GAC in chlorinated water, HVOCs desorbed from the carbon immediately after the contactor was put in service. This practice should be eliminated in a future full-scale treatment plant to avoid undesirable concentrations of the HVOCs occurring in the effluent.

# Lime Clarification Process Studies

## I. Lime-Sodium Hydroxide Single-stage Softening

The first process in the potable water Reuse Demonstration Plant treatment sequence was lime clarification. This process involved the addition of hydrated lime in a slurry to the rapid mixing basin. Ferric chloride was added in the rapid mix basin as a coagulant aid at a dosage of about 20 mg/L. Two-stage lime soda ash softening was tested at the Reuse Demonstration Plant. It was successful in softening the water but operational difficulties caused by scaling of the pumps needed to convey the water between the first- and second-stage of the softening process made this process unusable. Softening would have several benefits to both the ion exchange system and the RO system if it could be accomplished. Of the methods for softening that were evaluated, the lime-sodium hydroxide treatment provided the most promise.

The results of many laboratory and some plant-scale tests showed that single-stage combined lime-sodium hydroxide softening can be accomplished. The optimum procedure for hardness removal was to add 350 mg/L lime and then sodium hydroxide to maintain a pH of 11.6. Ferric chloride was used as a coagulant aid at a dosage of 20 mg/L. The cost of single-stage lime-sodium hydroxide softening was in the same range as traditional lime-soda ash softening. The cost of both softening methods was about 40% higher than lime treatment alone.

Increases in alkalinity and conductivity caused by the addition of sodium hydroxide created problems with reverse osmosis. Also, increased scaling potential was created in the ion exchange columns and the reverse osmosis system due to increased alkalinity. Due to these issues single-stage lime-sodium hydroxide softening on a continuous basis was not recommended.

## II. Effect of Aeration on Lime Dosage

The first treatment process in the potable reuse treatment sequence was lime clarification. Hydrated lime was added to the treated wastewater in a slurry to the rapid mix basin. The nominal flow of the undisinfected treated wastewater was 1 mgd. Ferric chloride was also added in the rapid mix as a coagulant aid at a typical dosage of 20 mg/L. Pilot studies had shown that the typical lime dosage to raise the pH to 11.0 was 400 to 500 mg/L. Disturbingly; during plant startup 600 mg/L was needed. The cause was determined to be free carbon dioxide dissolved in the water produced by the wastewater treatment plant pure oxygen activated sludge system.

Aeration was installed in the entry to the rapid mix basin where there was minimal contact time (3 to 4 minutes). This resulted in a 10% lime dosage reduction saving more than $10,000 per year at the Demonstration Plant. Conservative estimates predicted that a more effective aeration system installed in the raw water pump wet well would reduce lime dosage at least another 5% and over a period of four years this would save a total of more than $15,000. No additional aeration was installed, but the need was noted for a full-scale treatment plant. Savings due to raw water aeration in a 100 mgd full-scale treatment plant would be in excess of $1.5 million per year.

# Ammonia Removal Process Studies

## I. Ion Exchange for Ammonia Removal- Laboratory Study

Ion exchange employing the natural zeolite, *clinoptilolite,* was the primary method for removing ammonia in the reuse treatment process. The process design anticipated splitting the process flow between two ion exchange beds while the third was off-line being regenerated and backwashed. Once the ammonia concentration in the water leaving the bed reached a pre-described level it was removed from service. A regenerated standby bed then returned to service. Design conditions predicted that with a feed ammonia-N concentration of 22 mg/L the bed would be removed from service when the effluent concentration reached 1 mg/L. Under these conditions a 100 bed-volume service cycle was expected. The actual performance of the ion exchange process produced an ammonia level in the effluent almost four times the expected value.

Scaling of the ion exchange media by precipitated calcium carbonate became a problem in continuous service and appeared to be the principal cause of higher than expected ammonia concentration in the process effluent. The fouled media was effectively acid cleaned to restore near virgin ion exchange capacity with a material loss of about 5%.

Such fouling appeared to be the result of calcium carbonate super saturation in the feedwater and subsequent contact between the feedwater and the high pH regenerate brine. Improved process control mitigated this situation. Ammonia residual and potassium accumulation in the regenerate brine further affected ammonia removal levels by reducing the effective exchange capacity of the media. The system service cycle was modified to deal with these issues. The ion exchange system, even with these adjustments, could not meet the removal goal and produce water below 1 mg/L ammonia.

## II. *Ammonia Removal by Breakpoint Chlorination*

Secondary treated wastewater typically contains total ammonia nitrogen concentrations in excess of 20 mg/L as nitrogen. Water entering the Reuse Plant was typical and ammonia concentrations ranged between 15 and 35 mg/L. Alternatives to ion exchange for ammonia removal were investigated due to problems encountered with that system. Breakpoint chlorination became the first among these alternatives to be evaluated due to the availability and simplicity of the process.

Breakpoint chlorination will remove ammonia nitrogen from the reuse influent most optimally at a chlorine to ammonia weight ratio of about 8:1. The process though has several disadvantages. It creates chlorinated organic byproducts some of which are regulated. Increased concentrations of brominated volatile organic byproducts are also produced which may be effectively reduced by activated carbon treatment. Dechlorination would most likely be needed if breakpoint was practiced due to high chlorine residuals that would be present. Dechlorination could be accomplished by activated carbon with the added benefit of removing increased levels of halogenated organics. Dechlorination of monochloramine with activated carbon released ammonia back into the water. Breakpoint chlorination also produced increased chloride levels in the water. Depending on the amount of chlorine added, and the concentration in the untreated water, the chloride concentration may reach unacceptable levels affecting the reverse osmosis process that followed.

The cost of breakpoint chlorination was about $0.012 per million gallons. This cost was comparable to or lower than the selective ion exchange process tested at the Reuse Demonstration Plant. Due to the production of halogenated by-products breakpoint chlorination should only be used to polish the low level ammonia residual present in the finished plant product water. The formation of halogenated organic compounds prevents this process from use for removing high levels of ammonia.

## III. *Effect of Storage on Ammonia Concentration*

This study examined the effect of wastewater storage on ammonia concentration. The treated wastewater contained ammonia nitrogen (12 to 40 mg/L). Due to the expense and complexity of the ion exchange process alternatives for ammonia removal were evaluated. If ammonia removal could be achieved simply by storing the effluent this may be a possible method of reducing the overall treatment costs for ammonia.

Ammonia transformed first to nitrite and then to nitrate when the treated wastewater was stored under laboratory conditions. Diffusion of atmospheric oxygen into the sample provided needed oxygen even though the solution was unmixed. Aeration accelerated the process reducing the time required for conversion from ammonia to nitrate by about 50%. After more than 60 days of storage significant denitrification did not occur. The addition of ethanol to the nitrified sample brought about complete denitrification in only six days.

This method of ammonia removal, although "low tech," could be useful in a full-scale facility if the wastewater was stored before further treatment. Ammonia levels could be reduced or eliminated, thus, reducing the cost of further reduction.

# Water Quality Assessment Studies

## *I. Composite Sampler Development*

The water quality experts on the analytical and health effects advisory committee recommended using 24-hour composite samples for many of the water quality analyses. These samples would provide a representative assessment of water quality occurring at key points throughout the treatment process. After considerable investigation, commercial composite samplers for this purpose could not be located. The Reuse Plant staff consequently designed and constructed composite sampler's to be used in the Project.

These samplers were designed around small under-the-counter refrigerators fitted with microprocessor event controllers which activated Teflon© solenoid valves. The event controllers were programmed to start and stop on specific days, to control the number of sampling events, and sampling durations over a 24-hour period. A minimum of eight sampling events were used for a 24-hour composite. The duration for each event depended on the sample line pressures at each location. The temperature of all collected samples was maintained at 4°C during compositing. Samples were deposited in three different containers. The volatile organic compounds were collected in a specially designed zero-headspace container. Nonvolatile organics were collected in four liter glass bottles. Metals, radiological, and major ions were collected in a 6 L plastic carboy.

After collection, the sample splits were transferred to the appropriate sample bottles. An iced cooler kept the samples chilled when transporting samples from the Reuse Plant or the Denver drinking water treatment plant to the water quality laboratory. Transportation times were kept to a minimum in most cases less than 30 minutes.

The composite sampler's were tested extensively to verify that representative samples were collected, and that contaminants were not introduced by the materials of construction. In the sampling devices all the materials that came in contact with the water were Teflon©, stainless steel, or glass. Extensive testing of the zero-headspace sample collection container verified that volatile organic compounds were not lost during the sampling procedure. After thorough testing the composite samplers performed effectively.

## II. Treatment Plant Hydraulic Computer Model Development

A hydraulic computer model was developed as result of a suggestion by the project advisory committee. The computer model was capable of accounting for the effect plant tanks and dilution had on the concentration of substances in the water at various points through the reuse treatment process. The model was also used to assist in collecting samples when contaminants were introduced into the plant to assess the removal capability of each reuse treatment process.

A series of tracer tests were conducted to provide data necessary to refine and calibrate the hydraulic model. Two tracer substances were used for this purpose: sodium chloride and sodium bromide. The sodium bromide was preferred because of its low background concentration and it was unaffected by all unit processes except reverse osmosis which occurred late in the treatment train. Sodium chloride was also used because of its ease of use and the ability to acquire real-time data from on-line conductivity sensors.

The hydraulic model's value was evident when the initial results of the tracer tests identified an anomalous flow behavior at three process locations. Short-circuiting was occurring through the flocculation basins, the lime clarifier, and the ballast ponds. While none of these was considered critical to process performance, modifications were undertaken to remedy these problems.

To correct the short-circuiting in the flocculation basins, a deflector was installed at the inlet to the direct flow upward. The two intermediate baffles were rearranged to produce a better serpentine flow through the basin. These changes succeeded in improving the flow through the basins and increased the average detention time from 4 to 7 minutes.

An examination of the lime clarifier revealed that short circuiting was caused by the overflow weir not being level. An average deviation of about 1/4 inch was found. Leveling the overflow weir fixed this problem.

To correct the short-circuiting in the ballast ponds, air diffusers at the bottom of the basins were repositioned so they were perpendicular to the direction of the flow. The effect was significant but only increased the average travel time through the basin by 25%. A duct was installed at the inlet end of the pond to direct the flow upward but this did not result in any further improvement.

### III. Plant Contaminant Challenge Study

One of the most compelling studies performed at the Reuse Demonstration Plant was the introduction of high concentrations of a broad spectrum of contaminants to demonstrate the removal capability of the health effects treatment sequence. The selection of the test compounds for the Project included representative substances from all the major contaminant groups including organic chemicals, inorganic chemicals, radiological substances, and microbiological organisms.

The hydraulic computer model of the Reuse Demonstration Plant developed before the plant challenge study began was used to predict the time and the concentration that would be expected at key points throughout the process. The computer model was calibrated first using sodium chloride as a tracer and monitoring its concentration through the plant using conductivity. Sodium bromide was also used as a calibration chemical with its concentration determined by ion chromatography. The bromide proved to be an effective tracer used in many tests to calibrate and refine the hydraulic model.

The contaminant challenge study employed many compounds and substances added to the plant at very high levels (many thousands of times greater than experienced in wastewater or natural water samples). The contaminant spike was monitored through the major treatment plant processes. None of the substances added were detected in the plant effluent. Most substances were completely removed by lime clarification (the very first treatment process).

A special test was conducted after all the individual compound tests. This test included the preparation of a cocktail of many of the compounds used in the individual tests and they were added to the plant simultaneously. Again none of these contaminants regardless of the concentration were found in the plant effluent. The only substances removed by less than 50% by lime clarification were chloroform and chromium. The minute residual amounts of both of these substances were removed by carbon adsorption to near undetectable amounts and neither was detected in the plant effluent.

A total of 22 contaminants (Table 7-1) were added to the plant in concentrations often thousands of times higher (The last column that shows the amount added to the plant compared to the maximum found in untreated wastewater) than found in any natural water samples. The lime clarification treatment process and the carbon adsorption process showed the greatest impact on the removal of these contaminants. Processes downstream of carbon adsorption, such as reverse osmosis, then acted as final barriers to any traces that might remain. This study demonstrated that Reuse Plant treatment could eliminate chemicals and microbiological substances at levels far higher than those found in any water supply. Reuse plant treatment, thus, provided a level of safety never before achieved by conventional water treatment plants.

**Table 7-1**
**Substances Used for the Reuse Plant Contaminant Challenge Studies**

| Substance | Highest Concentration in Plant Influent (mg/L) | Concentration Introduced into the Reuse Plant (mg/L) | Challenge* Concentration (x influent Max) |
|---|---|---|---|
| Inorganic | | | |
| Arsenic | 0.003 | 34 | 11,000 |
| Chromium | 0.009 | 55 | 6,000 |
| Nitrate-N | 3.2 | 98 | 30 |
| Nitrite-N | 0.27 | 35 | 129 |
| Cyanide | <0.001 | 24.2 | 24,000 |
| Lead | 0.01 | 24.2 | 2,400 |
| Uranium | 0.08 | 6.8 | 85 |
| Organic | | | |
| Acetic Acid | <0.0001 | 5054 | $5 \times 10^7$ |
| Anisole | <0.0001 | 23 | $2 \times 10^5$ |
| Benzothiazole | <0.0001 | 86.2 | $8 \times 10^5$ |
| Chloroform | 0.0078 | 229.6 | 29,000 |
| Clofibric Acid | <0.0001 | 17.1 | $1.7 \times 10^5$ |
| Ethyl Benzene | <0.0001 | 25.1 | $2.6 \times 10^5$ |
| Ethyl Cinnamate | <0.0001 | 67.8 | $6.7 \times 10^5$ |
| Methoxychlor | <0.0001 | 44.6 | $4.4 \times 10^5$ |
| Methylene Chloride | 0.0171 | 230 | 13,000 |
| Tributyl Phosphate | <0.0001 | 69.4 | $6.9 \times 10^5$ |
| Gasoline (1st trial) | <0.0001 | 97.8 | $9.7 \times 10^5$ |
| Gasoline (2nd trial) | <0.0001 | 2115 | $2.1 \times 10^6$ |
| Particles | | | |
| Latex beads (3µ) | Not tested | $2 \times 10^6$ | >1000 |
| Virus | | | |
| Coliphage (JJ) pfu/100mL | $1.4 \times 10^5$ | $1.5 \times 10^7$ | 100 |
| Coliphage (MS-2) pfu/100mL | $4 \times 10^5$ | $8.6 \times 10^9$ | 40,000 |
| Polio (attenuated) | Not tested | $2.1 \times 10^{10}$ | >1,000 |
| Challenge Mixture | acetic acid, arsenic, benzothiazole, chloroform, chromium, ethyl cinnamate, nitrate, tributyl phosphate. All substances added in the same concentration as for the individual tests shown above. | | |

*The number of times the background amount that was added as a plant challenge.

# 8
# Plant Operation
# During Health Effects Study Period

During the process performance testing assessment period (October 1985-March 1988), the plant configuration was altered many times to examine the various unit processes. In addition, various special studies were conducted to evaluate treatment options and investigate treatment issues. The objective was to eliminate unreliable or unnecessary treatment steps without jeopardizing water quality or consumer safety. As a result of this analysis, a plant treatment configuration was adopted and used for the entire two-year whole-animal chronic toxicity and reproductive health effects study portion of the Project. Comparing the plant design to the health effects treatment configuration these important changes are noted:

- the elimination of one stage of activated carbon,
- the elimination of the selective ion exchange system and the ARRP process,
- the replacement of chlorine dioxide with ozonation as the primary disinfectant,
- chloramination (monochloramine) instead of chlorine dioxide as the residual disinfectant,
- And the addition of a parallel pilot-scale treatment sequence that substituted ultrafiltration for reverse osmosis.

The Direct Potable Water Reuse Demonstration Plant treatment sequence used during the two-year whole-animal health effects testing period (Figure 8-1) was:

1. High pH lime clarification
2. Recarbonation
3. Filtration
4. Ultraviolet Irradiation
5. Activated Carbon Adsorption
6. Reverse Osmosis (Or Ultrafiltration as a parallel pilot treatment sequence)
7. Air Stripping
8. Ozonation
9. Chloramination

The treatment processes used during the animal health effects study were selected after more than two years of plant operation and evaluation. The selected processes had demonstrated the capability of producing water meeting quality standards consistent with the Project goals. This section of the report contains descriptions of the operating experience for each of these treatment systems during the animal health effects testing period (January 1989-December 1990). Even though the processes had already been operated successfully, in some instances for several years, continued operation provided an opportunity to refine the operation and learn more that could be used when designing a future full-scale facility.

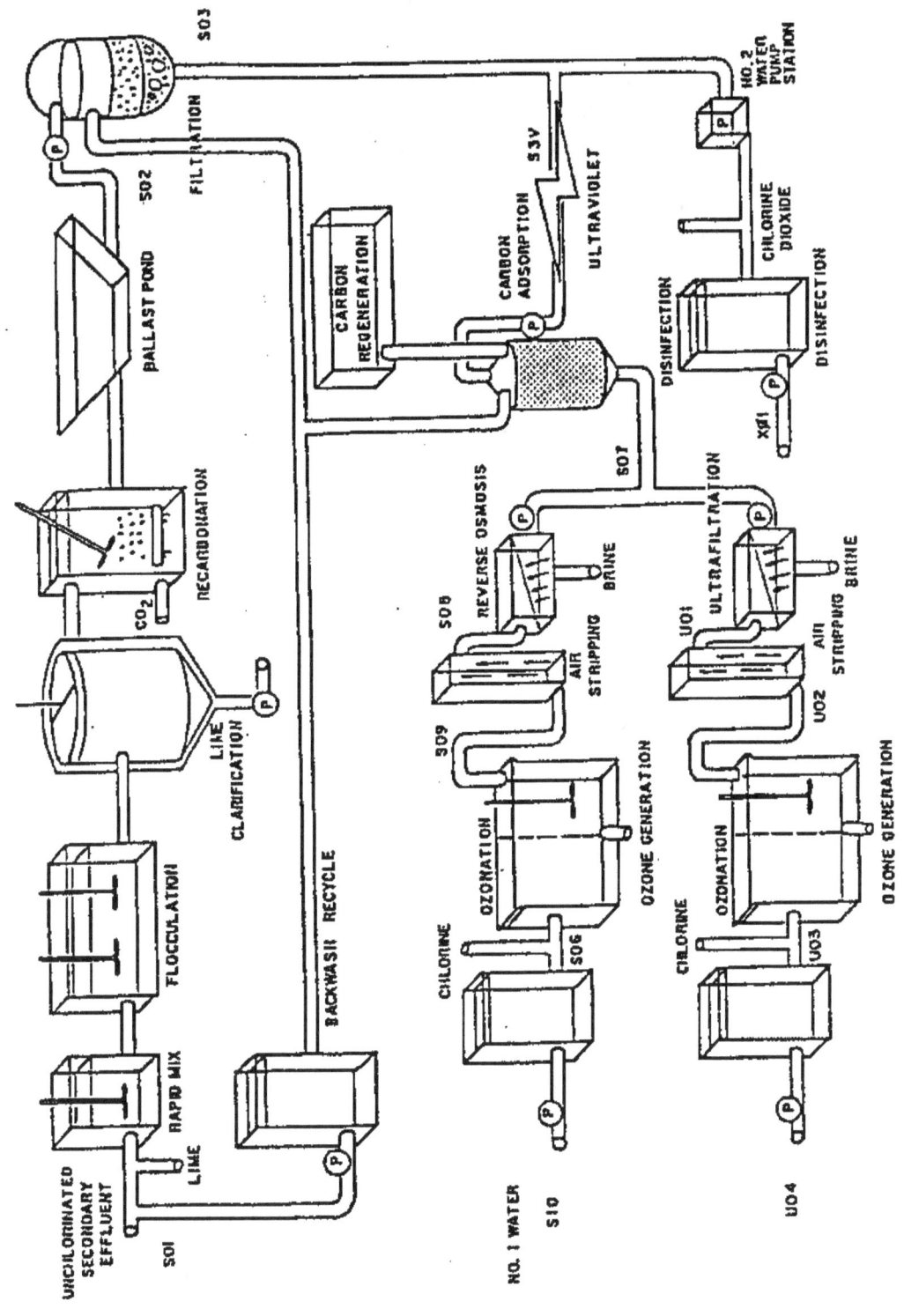

**Figure 8-1: Reuse Health Effects Treatment Sequence**

# Raw Water

| Flow Rate (average) | 0.90 mgd |
|---|---|
| Turbidity | 9.2 |
| pH | 7.0 |
| Sewage Treatment Plant Discharge Permit Requirements | 30 mg/L suspended solids 17 mg/L BOD$_5$ |

**Table 8-1:  Plant Influent Operating Parameters**

The raw water flow rate during the health effects testing phase was approximately 1 mgd (Table 8-1). Construction activities at the sewage treatment plant caused interruptions in raw water flow and periods of poor water quality. Most interruptions and upsets were of short duration and unplanned. Several lengthy interruptions, though, were predicted that warranted preventative steps to ensure continuous operation.

On several occasions construction at the wastewater plant required dewatering of the south effluent channel, which was the Reuse Plant source. When this took place the flow to the Reuse Plant raw water pump station was interrupted. During these times the chemical treatment processes (rapid mixing, flocculation, lime clarification, and recarbonation) were shut down until normal flows at the sewage treatment plant were restored. The processes downstream of chemical treatment continued operating from storage reserves in the Reuse Plant ballast ponds. Moreover, when the level in the wastewater plant channel was lowered large amounts of floating debris entered the raw water pump station wet well. On one occasion the raw water pumping suction had to be shut down so that the screen covering the mouth of the pump could be cleaned of debris.

The construction at the wastewater plant also required that portions of the wastewater aeration basin and south side secondary clarifiers be taken out of service. During these times the flow rate in the south effluent channel was lowered and the possibility of floating debris entering the raw the Reuse Plant raw water pump station also increased. Furthermore, during the low flow periods of early morning there was not enough water in the channel to sustain the raw water pump station operation. Because these conditions were expected to last for several months an alternative pumping system was needed.

The alternative pumping system used during the wastewater treatment plant construction consisted of a trailer mounted electric 1 mgd centrifugal pump. When it was needed, the pump was moved to a secondary clarifier at the wastewater plant and the suction hose placed within the clarifier near the overflow weir. The clarifier served as the pumping reservoir for the pump. An attempt to locate the suction piping in the effluent collector of the clarifier led to pump starvation problems and was discontinued. Discharge piping for the pump consisted of six inch flexible or PVC piping which was laid over ground and into the existing pump station where it was connected to the pump station piping at a pig launch point. The control system for the regular raw water pumps was modified to accommodate a third pump and provided remote control.

The trailer mounted pump was needed for two lengthy intervals. The first occasion was from April 18 to May 1, 1989. During this time the portable pump was used primarily during the low flow periods of early morning. On a second lengthier occasion the trailer mounted pump was used between June 22 and August 29, 1990. The trailer mounted pump was moved between two clarifiers to accommodate the construction. The south aeration basin was being modified during this time and this led to even lower flows through the wastewater treatment plant, this made the raw water pump station completely unusable. The portable pump then (for two months) successfully maintained treated wastewater delivery to the Reuse Plant. This was a critical necessity during the health effects study period since it ensured a continuous supply of reuse water concentrate samples for the ongoing animal health effects study.

As a matter of routine maintenance, the raw water pipeline was pigged (cleaned by swabs inserted into the pipeline) approximately twice per year. Additional pigging was required after periods of low flow in the south effluent channel at the sewage plant to remove debris that had settled in the line.

On several occasions high levels of organic carbon were detected in the wastewater. It was found that ethylene glycol used for deicing at the nearby international airport had entered the sewer plant. Total organic carbon concentrations exceeding 50 mg/L were measured in the Reuse Plant influent. Raw water surveillance thereafter facilitated shutting down the processes downstream of filtration until these short-term events passed.

# Chemical Treatment
## (rapid mixing, flocculation, lime clarification, recarbonation)

| Rapid Mix | |
|---|---|
| Flow Rate | 0.967 mgd |
| Detention time | 2.6 min |
| Velocity Gradient | 326/sec |
| Lime Dose | 471 mg/L |
| Ferric Chloride Dose | 20.9 mg/L |
| pH Setpoint | 11.0 |
| Flocculation | |
| Detention Time | 25 min |
| Velocity Gradient 1 | 125/sec |
| Velocity Gradient 2 | 100/sec |
| Velocity Gradient 3 | 100/sec |
| Chemical Clarifier | |
| Flow Rate | 0.967 mgd |
| Detention Time | 110 min |
| Surface Overflow Rate | 796 gpd/ft$^3$ |
| Waste Sludge Flow Rate | 0.022 mgd |
| Sludge Concentration | 5.2 wt% |
| Sludge Wasted | 9561 lb/day |
| Turbidity of Overflow | 1.3 NTU |
| Recarbonation | |
| Flow Rate | 0.945 mgd |
| Detention Time | 13 min |
| Velocity Gradient 1 | 533/sec |
| Velocity Gradient 2 | 188/sec |
| Carbon Dioxide Dose | 172 mg/L |
| pH Setpoint | 7.8 |

**Table 8-2: Lime Clarification Operating Parameters**

The flow through the rapid mix basin was approximately 1 mgd (Table 8-2). Detention time for rapid mixing was about three minutes and the lime dosage averaged 471 mg/L to achieve an 11.0 pH setpoint. Ferric chloride used as a coagulant aid was added at a dosage averaging about 20 mg/L.

The rapid mixers and pH probes were subject to scaling because of the high pH. During previous phases of operation it was determined that mixing efficiency declined when the mixer became scaled. Scaling severe enough to impact mixing took approximately 2 to 3 weeks. Based on this, the rapid mix basins were cleaned every two weeks. The pH probe used to control lime addition was cleaned with a diluted acid solution twice weekly as part of routine plant maintenance.

The average detention time in the lime clarifier was 82 minutes. Although this was less than the theoretical detention time of 110 minutes it was an improvement from less than 60 minutes before the clarifier overflow weir was leveled when the plant hydraulic computer model uncovered this problem. The surface loading in the clarifier was 763 gpd/ft$^2$. The mean sludge solids concentration was 5.2%. The effluent turbidity of the clarifier averaged 1.3 NTU. The mean flow rate in the recarbonation basin was again about 1 mgd and the mean pH was 7.7 using an average carbon dioxide dosage of 172 mg/L.

### Sludge handling

The primary method of sludge disposal was dewatering and landfill. Disposal by discharge to the sewage treatment plant was used as a backup when equipment failed or other problems were encountered. The dewatering method was used primarily to reduce cost.

A sludge vacuum filtration system was added as a process after plant startup. The unit was purchased used and did not have a backup. The equipment was plagued with many problems that resulted in considerable down time. While repairs were made the thickened sludge from the on-line clarifier was stored in the off-line clarifier. Additional thickening during storage improved the performance of the sludge filtration system.

Many of the problems with vacuum filtration system involved transporting the sludge cake from the drum to the container used for hauling. Originally a conveyor system was used. The conveyor was subject to mechanical failures and sludge buildup along the belt. The conveyor was replaced with a progressive cavity sludge pump. This improved the ability to move the sludge cake on into the hauling container but a problem with pump starvation resulted. Finally, the progressive cavity pump was fitted with a bridge breaker device and this eliminated most transport problems.

# Filtration

| | |
|---|---|
| Flow Rate | 0.945 mgd |
| Operating Pressure | 70 psig |
| Turbidity | 0.33 NTU |
| Hydraulic Loading Rate | 4.4 gpm/ft$^3$ |
| Average Filter Run Length | 22.9 hr |
| Length of Filter Backwash | 25 min |
| Backwash Flow Rate | 1500 gpm |
| Length of Surface Wash | 20 min |
| Surface Wash Flow Rate | 56 gpm |
| Backwash Loading Rate | 19 gpm/ft$^3$ |
| Terminal Pressure Drop | 11 ft |
| Filter to Waste | 10 min |

**Table 8-3: Filter Operating Parameters**

The average length of the filter runs was 22.9 hours (Table 8-3). Effluent turbidity averaged 0.33 NTU. The removal of microbiological contaminants and total organic carbon across the filters was modest. But the removal of these contaminants was not the primary function of the filters.

A loss of dissolved oxygen across the filters, approximately 42%, and nitrite in the filter effluent indicated the growth of nitrifying bacteria within the filters. The dissolved oxygen loss was controlled by disinfecting the filters with chlorine dioxide. As a result, process control testing included measuring nitrite concentrations across the filters. A standard operating procedure was developed for disinfecting the filter's off-line with chlorine dioxide approximately twice per month.

# Ultraviolet Irradiation (UV)

| Flow Rate | 0.082 mgd |
|---|---|
| UV Lamps | Ten 60 watt mercury vapor at 254 nm |
| Contact Time | 13 sec |
| Cleaning Solution | 2 wt% citric acid |

**Table 8-4: UV Operating Parameters**

The UV system (Table 8-4) proved effective in reducing coliform bacteria populations in the flow stream. Overall bacteria populations (mHPC) were less affected by this treatment process but still removal was significant (coliphage virus inactivation was greater than 99%).

The manufacturer recommended lamp replacement after 7,500 hours. Performance remained consistent so lamp replacement was extended to more than 14,000 hours of service. At this point coliform counts in the UV effluent began to rise prompting the replacement of the lamps. Microbial reduction results returned to the expected levels after lamp replacement.

The UV system was cleaned when the intensity of the UV light dropped by 25%. Cleaning was required more often with older lamps. With the old lamps cleaning was required approximately once per month. With the new lamps the time between cleanings was extended to several months.

Although the UV system was effective in reducing bacterial population in the flow stream, there was no evidence that it had an effect on the performance of the following processes. Data collected from these processes showed that coliform counts did not change through carbon or reverse osmosis and it appeared that growth did occur again in air stripping tank.

The UV system was inserted into the treatment configuration in an attempt to reduce bacterial loading on the carbon columns that followed. Evaluations conducted during the health effects treatment period suggested that the UV system was of limited value as an intermediate process and probably would not be included in design of a future full-scale facility.

# Activated Carbon

| Flow Rate | 0.082 mgd |
|---|---|
| Contact Rate | 1.37 BV/hr |
| Hydraulic Loading Rate | 4.5 gpm/ft$^3$ |
| Empty Bed Contact Time | 42 min |
| Length of Backwash | 20 min |
| Backwash Flow Rate | 200 gpm |
| Backwash Loading Rate | 15.9 gpm/ft$^3$ |
| Terminal Pressure Drop | 15 ft |
| Effluent Turbidity | 0.25 NTU |
| Post-backwash Filter to Waste | 20 min |
| Carbon Media | 8x30 mesh Filtrasorb® 300, 26'8" depth |

**Table 8-5: Activated Carbon Operating Parameters**

The carbon columns (Table 8-5) were hydraulically loaded at 4.5 gpm/ft$^2$ or approximately 86% of the design loading rate. The hydraulic throughput rate was 1.37 bed volumes per hour and the empty bed contact time was 42 minutes. Backwashing was performed at 15.9 gpm/ft$^2$ for 20 minutes.

The lower loading rate on the activated carbon adsorption process was a result of the increased pressure headloss caused by biological growth within the carbon column and the inability (caused by a design accommodation) to effectively backwash the carbon bed. Although this combination led to shorter service cycles, it did not appear to impact the contaminant removal characteristics of this process.

Backwashing problems resulted from a large headspace in the contactor above the carbon bed. This configuration was a design concession to accommodate the option of operating the process in either up-flow or down-flow mode. As a result particulates fluidized during the backwash process were not effectively removed from the contactor and settled on the surface of the bed at the conclusion of the backwash. This led to the formation of a mat of carbon fines combined with biomass at the top of the filter bed that severely restricted flow through the bed. Two procedures were employed to combat the flow restriction problem: the terminal head loss set point was increased to 35 feet; and the top of the carbon filter was periodically scraped. Even with these actions the system could only be operated at the reduced loading rates.

One of the difficulties encountered in operating the carbon columns was the formation of hydrogen sulfide within the columns. The formation of the gas within the carbon contactor created not only the odor problems but increased the potential for corrosion and other problems downstream. A review of the scientific literature revealed that the addition of sodium nitrate to the feedwater of the carbon column could reduce hydrogen sulfide production. Bacteria populations would preferentially reduce nitrate to nitrogen gas over the reduction of sulfate to hydrogen sulfide gas. Adding a regulated foreign substance such as sodium nitrate to the flow stream was unacceptable particularly during the health effects study.

Nitrite was used to combat this problem, but it was added by simply changing conditions within the plant. The concentration of nitrite in the carbon influent flow was increased or decreased by adjusting the aeration in the ballast pond ahead of the filters. When aeration was taking place the nitrite concentration in the filter effluent which was the carbon system influent would increase. When the aeration was turned off nitrite levels would then decrease. Although the concentration could not be controlled precisely, it was easily influenced up or down using this procedure.

Changing the nitrite concentrations in the carbon feedwater by this procedure produced results that matched those reported by other researchers. When the nitrite concentrations increased, the hydrogen sulfide concentrations decreased. Although no attempt was made to quantify the reductions in hydrogen sulfide concentration, the odor of the gas disappeared. The conditions were changed several times with identical results obtained each time. This procedure eliminated the hydrogen sulfide nuisance gas.

The TOC across the carbon column was monitored as a control parameter. When new or regenerated carbon was added to the carbon beds high levels of TOC removal were experienced. This performance tapered off quickly to a relatively steady state of about 48% removal for most of the health effects treatment period. Biological activity within the bed appeared to be responsible for this performance. Thus, carbon regeneration was unnecessary for lengthy periods since a 48% removal was adequate for the reuse treatment because membrane treatment followed. The carbon columns used during the health effects study were operated for more than two years without regeneration and the performance remained at about 50% removal during that entire time. This operational result could eliminate the need to include a carbon regeneration furnace in a future full-scale treatment plant.

# Reverse Osmosis

| Feed Flow Rate | 0.082 mgd |
|---|---|
| Feed Pressure | 359 psig |
| Feed Conductivity | 1072 μmhos/cm |
| Feed Turbidity | 0.22 NTU |
| Product Conductivity | 69.4 μmhos/cm |
| Product Water Recovery | 86.3% |
| Rejection (conductivity) | 93.3% |
| Hydrochloric Acid Dose | 121 mg/L |
| System Components | Three Units, 4-2-1 array with 7 vessels each |
| Membranes | Fluid Systems® model 4600 thin film composite, spiral wound polyamide |
| Cartridge pre-filters | 5μm |
| Scale Inhibitor | Sodium Hexametaphosphate, 5 mg/L |

**Table 8-6: Reverse Osmosis Operating Parameters**

The feed flow rate through the reverse osmosis system (Table 8-6) averaged about 0.1 mgd during the health effects study period. The mean specific conductance of the product water was 69 μohms/cm. The average rejection based on specific conductance was 93.3% and the mean recovery was 86.6%, hydrochloric acid was added to the feed flow at a dosage of 121 mg/L, and sodium hexametaphosphate was added at a dosage of approximately 5 mg/L.

During the first month of the health effects study operation a substitute scale inhibitor, Flowcon® 100, was tested. After evaluating system performance and comparing the cost it was determined that there was no advantage to using this scale inhibitor. Sodium hexametaphosphate was then used during the rest of the study.

Midway through the health effects study the elements were replaced in some of the RO pressure vessels. This was done because the elements were approaching the end of their expected life and performance had declined. Leaving old elements in one of the units provided a basis for comparing the new elements with the old.

The new elements were installed in units one and three and as expected the feed pressure required to operate these units was significantly less than that required to operate the older elements. Also, as expected, the product water quality was improved with the newer elements.

The performance of the new elements was far superior to the performance of the old elements. In a full-scale facility an evaluation would be needed to determine the most cost-effective time to replace the elements. This type of evaluation was not performed during the Demonstration Project. It should be noted that the new elements were not operated at the design flow of 35 gallons per minute. Instead a feed flow of 30 gallons per minute was used. Attempts to operate at the higher feed flow were hampered by the extensive head loss in the carbon columns that preceded reverse osmosis.

Cleaning of the RO unit was initiated when the normalized feed pressure increased by 10%. The normalization procedure involved entering data into a manufacturer supplied computer program and interpreting the output. The inexperience of some operations staff and uncertainty in using the computer program led to the development of a more straight-forward spreadsheet program that also normalized the data. Although the spreadsheet program was not as sophisticated as the manufacturer supplied program, it was shown to provide similar results. Instituting use of the spreadsheet program led to a more uniform cleaning program by taking into account performance differences due to temperature or conductivity changes in the feedwater. Cleaning was based more on fouling than perceived changes in performance that might have been due to temperature changes or dissolved solids loading.

The product water leaving the RO system was consistently of high quality as measured by specific conductance. The removal of specific contaminants by RO was thoroughly evaluated during the process assessment phases.

The silt density index test (SDI) was supposed to predict fouling potential for the RO system but this proved to be inaccurate. The values obtained by this testing regularly exceeded the manufacturer's recommended value for RO feedwater. In fact, nearly every SDI test taken from the beginning of the Project produced a result that would indicate that the system should be shut down but it operated successfully. SDI testing was terminated because it had no value.

In a separate evaluation the effect of operating the reverse osmosis system at 95% recovery (see Chapter 7) was examined. The study was performed because of the difficulty in disposing of the brine waste in the Denver area. By increasing the process recovery the volume of brine would be reduced by half. This study demonstrated that the reverse osmosis system could be successfully operated at this level and that a full-scale Reuse Plant should strongly consider using this mode of RO operation.

# Air Stripping

| Flow Rate | 0.079 mgd |
|---|---|
| Gas/Liquid Ratio | 100:1 |
| Column Packing | Polyethylene Tri-pack 3" |

**Table 8-7: Air Stripping Operating Parameters**

The influent and effluent pH values for the air stripper (Table 8-7) flow averaged 4.8 and 6.4 respectively. On average there was a net increase in both the coliform bacteria concentrations and membrane heterotrophic plate count (mHPC) concentrations across the air stripper. The air stripper was routinely disinfected with chlorine dioxide to control bacteria growths. The air stripper was effective in removing hydrogen sulfide and volatile organic compounds (Chapter 7) from the flow stream when present.

# Ozonation

| Flow Rate | 0.079 mgd |
|---|---|
| Detention Time | 86.5 min |
| Ozone Residual | 0.14 mg/L |
| Ozone Off-gas Concentration | 0.028 wt% |
| Applied Ozone Dose | 0.67 mg/L |
| Absorbed Dose | 0.62 mg/L |
| Generator Power Consumption | 4.06 kwh/day |
| System Components | Two Generators, Air Dryer, Catalytic Off-Gas Ozone Destruction |

**Table 8-8: Ozonation Operating Parameters**

The applied ozone dosage (Table 8-8) was 0.67 mg/L and the absorbed dosage was 0.62 mg/L. This resulted in a mean transfer efficiency of over 92%. The detention time in the ozone contractor was 86.5 minutes and the average residual at the end of the contractor was 0.14 mg/L. Power consumption for the generation of ozone averaged 4.06 kW per day. The mean membrane heterotrophic plate count concentration in the effluent was approximately 16 per milliliter. Coliform bacteria and coliphage viruses were not detected in any samples after ozone treatment.

# Chloramination

| Chlorine Dose | 0.97 mg/L |
|---|---|
| Total Chlorine Residual | 0.56 mg/L |
| Chlorine Contact Time | 14.6 min |
| Turbidity | 0.06 NTU |

**Table 8-9: Chloramination Operating Parameters**

The average chlorine dose (Table 8-9) during the health effects phase was 0.97 mg/L. The average residual was 0.56 mg/L total chlorine and the contact time was 14.6 minutes.

# Ultrafiltration Pilot-size Treatment (UF)

| Ultrafiltration Membranes (3) | Desalination Systems® Model G-10 Thin Film Polysulfone, Molecular weight cutoff 2500 |
|---|---|
| Flow Rate/unit | 0.43 gpm |
| Recovery Rate | 80.5% |
| Feed Pressure (avg) | 97 psig |

**Table 8-10: UF Operating Parameters**

At the start of the health effects study, the ultrafiltration system (Table 8-10) consisted of two parallel units. A third unit was added during the study to increase the pilot system capacity. The average recoveries for the units were 78.5% to 81.1%. Feed pressure varied depending on the degree of fouling but was as low as 60 psig for new elements and as high as 175 psig for older elements.

The ultrafiltration system was cleaned approximately once per month. Midway between cleaning cycles the system was removed from service and flushed at low pressure. The low-pressure flush was effective in temporarily reducing the feed pressure and this was consistent with manufacturer's suggested operation.

The pilot-size ozone contactor reduced the membrane heterotrophic plate count by approximately a factor of ten. The mean coliform count and the concentrations of coliphage in the ultrafiltration side stream influent were below detection limit and so quantifying removals of these parameters was not possible.

Several problems developed for pilot plant ozone system. The first of these involved the loss of ozone residual whenever the backup carbon column was put into service. This was caused by hydrogen sulfide in the backup column effluent. The same problem was not seen in the main treatment plant ozone contactor following reverse osmosis.

A second problem that occurred with the pilot-sized plant was a turbidity increase across the ozone contactor. It was found that the turbidity was coming from the air carrying the ozone. Attempts to filter the airstream were ineffective. Again this problem was not seen in the main plant flow stream.

Chlorine contact tank used in the pilot-sized ultrafiltration treatment system suffered from short-circuiting and did not provide optimum time for disinfection. The membrane heterotrophic plate count increased by a factor of two through the tank. Attempts to improve chlorine addition and control were unsuccessful.

The pilot-size ultrafiltration treatment plant provided a means of comparing the treatment provided by an ultrafiltration process with that provided by a reverse osmosis process. The results corroborate the expectation that the reverse osmosis system had superior removal characteristics compared to the ultrafiltration system. Of particular interest was the organic removal in the ultrafiltration pilot system. The ultrafiltration units removed 80.5% of the total organic carbon and 29.6% of total organic halogen. This illustrates the ultrafiltration effectiveness for organics removal and by inference the precursors for chlorinated byproducts formation. The ultrafiltration system operated continuously for the entire health effects study period and provided a reliable product that was tested along with the main plant product water.

## Conclusions

The reuse treatment plant was operated continuously during the two-year animal health effects study. All treatment processes ran optimally producing water meeting the quality goals for the Project. Although the treatment system was complex, the instrumentation and controls coupled with intensive operator attentiveness ensured superior process performance. Further, continuous superior plant operation confirmed that the plant could be operated by the same number of operators with the same qualifications as conventional water treatment plants. These results along with operating experience from the earlier three-year plant process optimization operational period demonstrated the unquestioned reliability of the plant thus meeting the Project goal for this factor.

# 9

# Cost Estimates

The cost of the extensive treatment required to ensure the safety of direct potable water reuse was estimated shortly after the plant began process evaluations. These estimates were based upon the treatment system designed into the demonstration plant. Although the early estimates included the economy of scale for a projected 100 mgd treatment plant, they only used actual operating experience from a brief period. Initial estimates indicated that the cost for a projected full-scale treatment plant operation was comparable to the estimated cost of proposed future traditional water resource supply projects. To refine the cost estimates information was continuously collected while the treatment plant was in operation.

The revised cost estimates shown in this section were based on the health effects treatment system processes and actual operating experience for at least two years. Additionally, the estimates for traditional water supply resource projects were based upon the exhaustive Army Corps of Engineers *Environmental Impact Study* (May 1988) that included an assessment of Metropolitan Denver's water supply alternatives. This extensive review evaluated the cost of various future projects that could be used to satisfy Denver's projected water demands. These projections spanned the same time frame as was expected for possible implementation of direct potable reuse. The *Environmental Impact Study* identified several future projects with costs ranging from $250 per acre foot to $960 per acre foot. The direct potable reuse treatment cost projections needed to fall within this range to be competitive.

Before cost estimates could be prepared, a treatment scheme for the future full-scale facility needed to be identified. Based on work completed at the demonstration facility the selective ion exchange ammonia removal treatment process was not included in the full-scale facility cost estimate. The Metropolitan Wastewater Reclamation District plant was already making plans to add partial nitrification treatment. Based on this knowledge a determination was made that the full-scale reuse facility would include a biological denitrification step to complete the nitrogen removal sequence for the reuse process water. This process was included as the first treatment at the projected full-scale potable reuse treatment plant.

Using the performance data obtained from the various process assessment phase evaluations the treatment systems following nitrogen removal at the future full-scale facility would include high pH clarification, filtration, carbon adsorption, and the associated side stream processes necessary for each of these systems. These treatment processes would then be followed by one of two options: either complete treatment of the process flow stream by reverse osmosis or a blended flow stream treated half by reverse osmosis (RO) and half by ultrafiltration (UF). These membrane processes would then be disinfected with ozone and chloramines would be used as a secondary residual disinfectant.

The second treatment sequence using a 50/50 blend of RO and UF was considered for several reasons. The main drawback in considering total flow RO treatment for a reuse supply was the disposal of a large quantity of brine or reject water. An advantage of a split treatment using a UF process, which does not reduce salt levels, was the reduction in brine waste produced. The RO process must be used to treat at least half of the flow in the full-scale facility to meet the water quality requirements, such as total dissolved solids, imposed on the Project. A second benefit realized by a split RO and UF treatment plant was the fact that the water treated strictly by RO was extremely corrosive and would require additional treatment while the blended flow would not need this extra stabilization.

Since both of the options (RO/UF split treatment and the entire flow treated by RO) were possible, feasibility cost estimates were compiled for both. Either treatment train would require disinfection of the final process water before distribution. To accomplish this disinfection the full-scale treatment plant would include ozonation as the primary disinfectant with chloramination as the secondary residual disinfectant. The full-scale reuse facility would thus contain the following treatment processes: nitrification/denitrification, high pH clarification, filtration, carbon adsorption, reverse osmosis or a RO and UF combination, air stripping, ozonation, chloramination, and the associated support processes necessary for each main unit process.

Many of the processes chosen for the health effects study treatment sequence were not typical water treatment plant processes. For this reason, several references were needed to obtain much of the cost information. Construction cost estimates for some of the processes were calculated using engineering table references. Operational and maintenance costs were derived from the actual operating experience gained from the Reuse Demonstration Plant.

The location of the future full-scale reuse facility had not been determined. For this reason the cost estimates for piping necessary to connect the reuse facility with the wastewater treatment plant and to the distribution system were not included. But, so that the cost estimates were representative, several assumptions were made including:

- To ensure that the raw water flow would have sufficient pressure within the plant a pumping facility capable of delivering 30 feet of total dynamic head was included in the high pH clarification system cost estimate.

- The exact method of brine disposal also had not been determined. The cost for disposing of brine could greatly impact the total cost of potable water reuse. An evaporation pond system was included in the cost estimate since this method had been employed elsewhere and cost estimate could be calculated.

- In developing the treatment cost estimates an assumption was made that the distribution system necessary to deliver the finished water to the customer was already in existence.

A final point which must be considered when reviewing the cost estimates was that the treatment facility was assumed to be in operation 100% of the time. Most facilities do not operate at 100% load factor and only reach full operating capacity infrequently. This assumption increased the operational cost thus resulting in a higher total.

The actual operation and maintenance costs for the Reuse Demonstration Plant were higher than the estimates obtained from engineering tables. These actual reuse operational costs were used to develop conservative cost estimates. In the case of sludge removal, the Reuse Plant was dependent on other companies for disposal which affected the cost. For a full-scale facility sludge handling from the plant included a re-calcining furnace to recover used lime that would reduce this cost significantly. Carbon regeneration costs for the Reuse Plant were derived from operating the furnace during a six month period. The regeneration costs obtained from engineering tables were calculated on the basis of operating the furnace twice a year. Operating experience, however, suggested that regeneration might not be required for two years or longer thus reducing the cost. It may even be more cost-effective to purchase regenerated or new carbon rather than to operate a carbon regeneration furnace infrequently on site.

Additionally, the cost estimates, in some cases, were based on small scale operating experience such as ultrafiltration. The full-scale plant operational cost should benefit from an efficiency of scale. Other factors which may reduce the costs include Metropolitan Wastewater Reclamation District future treatment that may include denitrification which will eliminate the need for this process in the full-scale facility.

Ultraviolet radiation was not included in the 100 mgd reuse facility cost estimate. From experience gained during the health effects treatment period. It was determined that it would not be needed in a full-scale facility.

By substituting the higher cost estimates from the Reuse Demonstration Plant operating experience with available capital construction costs estimates and upper cost limit was calculated of $2.33 per thousand gallons (Table 9-1).

The cost estimates calculated for a full-scale facility utilizing engineering tables for capital construction and actual Reuse Demonstration Plant values for process operation and maintenance a range from $534 per acre foot to $762 per acre foot. These estimates compare favorably with those of equal uncertainty for future water supply augmentation projects described in the final *Environmental Impact Study* that evaluated conventional Denver water supply alternatives. These estimated ranged from $250 per acre foot to $960 per acre foot. Cost, therefore, does not appear to be a barrier to future implementation of direct potable water reuse in Denver.

**Table 9-1**
**Cost Estimates for 100 mgd Potable Water Reuse Treatment Plant**
($/1000 gal all costs in 1991 dollars)
(100% RO Treatment Plant, 50% RO and 50% UF Treatment Plant)

| Treatment Process | RO Treatment Plant | | | 50%RO and 50% UF Treatment Plant | | |
|---|---|---|---|---|---|---|
| | Amortized Capital | O & M | Total | Amortized Capital | O & M | Total |
| Biological Denitrification | 0.09 | 0.05 | 0.14 | 0.09 | 0.05 | 0.14 |
| High pH Lime Clarification | 0.15 | 0.42 | 0.57 | 0.15 | 0.42 | 0.57 |
| Filtration | 0.03 | 0.02 | 0.05 | 0.03 | 0.02 | 0.05 |
| Activated Carbon | 0.09 | 0.19 | 0.28 | 0.09 | 0.19 | 0.28 |
| Reverse Osmosis | 0.46 | 0.72 | 1.18 | 0.21 | 0.39 | 0.60 |
| Ultrafiltration | -- | -- | -- | 0.10 | 0.17 | 0.37 |
| Air Stripping | 0.0015 | 0.0008 | 0.0023 | 0.0015 | 0.0008 | 0.0023 |
| Ozonation | 0.008 | 0.07 | 0.08 | 0.008 | 0.07 | 0.08 |
| Chloramination | 0.002 | 0.003 | 0.005 | 0.002 | 0.003 | 0.005 |
| Miscellaneous | 0.01 | 0.017 | 0.027 | 0.01 | 0.017 | 0.027 |
| **Grand Total** | **0.84** | **1.49** | **2.33** | **0.69** | **1.43** | **2.12** |

# 10

# Individual Process Contaminant Removal

The Reuse Demonstration Plant treatment system employed a sequence of individual processes (Figure 3-1) linked in series to produce water of potable quality. The plant design included these separate processes: lime clarification, recarbonation, filtration, ion exchange, activated carbon (first-stage), ozone, activated carbon (second-stage), reverse osmosis, air stripping, and chlorine dioxide disinfection. Before starting the animal health effects study ultraviolet irradiation, ultrafiltration, and chloramination were added (Figure 8-1).

Each unit treatment process was continually examined to reach optimum performance. Chapter 6 discussed the operational and performance elements of each process that affected its selection or rejection for use during the animal health effects study period. These process settings affected water quality produced by each process, its cost of operation, and the process reliability. The contaminant removal results discussion was narrowly focused on the primary function of each unit process. This chapter broadens that discussion to examine the contaminant removal performance for a broader range of pollutants for each of the primary treatment processes.

## Process Performance Monitoring Results

In addition to the many process control water quality tests performed hourly by the operators (like pH, turbidity, nitrite, dissolved oxygen, specific conductance, ammonia, % sludge solids, chemical delivery screening tests, and dosage optimization jar tests) a suite of tests were used to further characterize the performance of the treatment processes. This group of tests were performed mostly on a weekly basis (coliform and standard plate count were tested daily) by technicians at the Department's Water Quality Control Laboratory. Mean test results for these analyses are shown in Table 10-1.

**Table 10-1**
**Individual Process Performance Monitoring Results**
% Removal (average of mean results from all phases)

| Contaminant | Lime | Filter | IE | UV | Carbon | RO | UF | O$_3$ |
|---|---|---|---|---|---|---|---|---|
| Total Coliform | 99.9 | 38.7 | 0* | 99.99 | 31.0 | 98.6 | 100* | 99.9 |
| Coliphage 137 or C | 98.7 | 23.5 | 6.6 | 99.99 | 31.3 | 100** | 100* | 99.9 |
| Coliphage B | 99.6 | 0* | 3.3 | 99.5 | 71.2 | 100** | 100* | 100** |
| mHPC | 15.6 | 24.6 | 50.0 | 98.7 | 29.5 | 99.99 | 97.9 | 97.0 |
| Total Organic Carbon | 44.5 | 7.6 | 5.7 | -- | 63.6 | 80.0 | 78.2 | 2.5 |
| Total Alkalinity | 10.2 | 2.7 | 0* | -- | 0* | 98.7 | 22.6 | 0* |
| Hardness | 13.6 | 0* | 37.7 | -- | 0* | 98.9 | 28.6 | 0* |
| Turbidity | 83.3 | 78.0 | 12.5 | -- | 42.4 | 64.7 | 80.0 | 0* |
| Ammonia | 1.6 | 0* | 91.3 | -- | 8.3 | 88.5 | 27.1 | 4.5 |
| Specific Conductance | 0* | 0* | 0.9 | -- | 0* | 93.5 | 29.9 | 0* |

Lime = lime clarification; filter = filtration; IE = ion exchange; UV = ultraviolet irradiation; carbon = activated carbon; RO = reverse osmosis; UF = ultrafiltration; O$_3$ = ozone
\* not removed or not significantly reduced
\*\* none detected in process effluent
-- not measured

The results shown in the table are from several treatment plant operational periods. Some processes were operated only during certain times and may have been positioned so that the preceding processes may have removed some contaminants below the analysis detection limit. This would make calculating its removal percentage impossible. During another operating period the process contaminant removal could be calculated. So, caution is called for when examining the results in this table. It is best to use these results as an indication of the removal performance rather than to rely on the exact values.

For example Carbon (Activated Carbon) is shown to remove about 30% of the Total Coliform, Coliphage 137, and Standard Plate Count. Coliphage B removal is even higher at more than 70%. These removals seem impressive. And they may be. But, Carbon was operated both in single and two-stage modes. The second-stage was preceded by ozone and later UV so the number of bacteria and coliphage entering the column was low. Removal of these contaminants then may have been impacted by the amount that was present in the process feed. According to the results presented in the table Carbon did remove these microbiological organisms but the primarily function is to remove organic carbon compounds and those substances should be the focus of an analysis of the performance of this process.

The process performance monitoring suite of test parameters included five main categories of contaminants: (1) Particles, (2) Microbiological organisms, (3) Organic Carbon, (4) Ammonia, and (5) Minerals. Removal of contaminants in these categories is discussed for each of the individual treatment processes.

### *Particle Removal*

Turbidity was the analysis measure used in the process performance monitoring test suite as an indication of particle removal. The primary particle removal processes were lime clarification (83.3%), filtration (78.0%), reverse osmosis (64.7%), and ultrafiltration (80.0%). The apparent low value for removal by reverse osmosis is probably a consequence of measurements at the limit of detection. Activated carbon reduced turbidity by 42.4%. This makes sense because it is a filter-like process even though the primary function of this process is organic carbon removal.

### *Microbiological Organism Removal*

Microbiological organisms are removed or inactivated by several processes. Analyses used for this performance measure were: total coliform, standard plate count, coliphage 137, and coliphage B. Based on total coliform (results were similar for the other microbiological contaminants tested) the main processes removing microbiological organisms were lime clarification (99.9%), ultraviolet irradiation (99.99%), reverse osmosis (98.6%), ultrafiltration (100%), and ozone (99.9%). The relatively low removal value for reverse osmosis was again an artifact caused by the infrequent detection of coliform during an operational period when the system was inoculated with organisms and equipment couldn't be disinfected effectively. Once this was corrected no microbiological organisms were detected in the reverse osmosis product water.

### *Organic Carbon Removal*

Total organic carbon was used as a performance monitoring parameter to evaluate the removal of organic substances by the Reuse Plant treatment processes. The primary organic carbon removal processes were lime clarification (44.5%), activated carbon (63.6%), reverse osmosis (80.0%), and ultrafiltration (78.2%). Activated carbon removed this performance measure at high levels (>80%) for a short period when using virgin or regenerated carbon. This removal declined gradually to reach a steady level of about 50% after three to five months of operation. The air stripping process was not included in the testing of the performance monitoring parameters since it was considered to be part of the reverse osmosis system. Separate testing confirmed that total organic carbon was not significantly reduced by this process but individual volatile organic compounds were effectively removed.

### Ammonia Removal

Ammonia is removed by ion exchange (91.3%) and reverse osmosis (88.5%). Ion exchange was not used during the animal health effects study so reverse osmosis was the primary removal process during this operational period. The ultrafiltration process removed 27.1% of the ammonia present. These removals were adequate for Reuse Plant testing and to provide samples for the animal health effects study.

### Mineral Removal

Process performance monitoring parameters used for minerals were hardness, alkalinity, and specific conductance. All these measures showed similar results. Reverse osmosis was the only process designed to minerals (98.9% hardness reduction). This expected result confirmed one of the primary reasons reverse osmosis was included in the plant process design. Lowering the water mineral content was necessary to produce water comparable to Denver drinking water. Ultrafiltration surprisingly removed more than 28% of the hardness and lowered specific conductance by a similar amount. Lime clarification was tested in both single and two-stage softening modes. Hardness was significantly reduced when operating in the softening mode but resulting sulfate and chloride increases caused problems with reverse osmosis. The plant operational design also did not facilitate operation in a two-stage softening mode.

## Lime Clarification Process Contaminant Removal Assessment

The lime clarification process consisted of rapid mix, flocculation, clarification, and recarbonation. During the health effects treatment period lime was added at the rapid mix to reach a pH setpoint of 11.0. Ferric chloride (20 mg/L) was added as a coagulation aid. Settling occurred in the solids contact clarifier. The clarified water pH was then adjusted (7.8) by carbon dioxide in the recarbonation process.

Pilot testing at the Water Department's research facility and extensive literature references had predicted the capability of this process to remove a broad array of potential water contaminants. The removal capability for separate contaminants could only be established for those that were present in the feed water. To augment these results the plant-scale contaminant removal challenge study (Chapter 7) was conducted to confirm these predictions. Table 10-2 contains a list of the contaminants added (at extremely high concentrations) during this study to evaluate the removal capability of the plant to withstand an unusual pollution event. Table 10-2 shows the contaminant removal results for the lime clarification process from the contaminant challenge study.

Most contaminants are either completely removed or significantly reduced by lime clarification treatment. None of the microbiological organisms survived this treatment process. Most metals and organic compounds were similarly reduced. Of the inorganic substances only cyanide and chromium weren't reduced by more than 50%. Chloroform, methylene chloride, and clofibric acid were the only organic compounds removed by less than 50%. Clofibric acid was a substance known to resist conventional wastewater treatment, and it also was not removed by lime clarification. This compound was detected after activated carbon treatment.

These results substantiated the value of lime clarification as the leading process in the reuse treatment sequence. A broad array of contaminants were removed or substantially reduced and this product provided improved feed water for the remaining treatment processes.

**Table 10-2**
**Lime Clarification Process**
**Contaminant Challenge Special Study Results**

| Contaminant | % Removal |
|---|---|
| Lead | 100* |
| Chromium | 44.3 |
| Cyanide | 29.3 |
| Uranium | 79.8 |
| Arsenic | 70.0 |
| Nitrate | 100* |
| Acetic Acid | 100* |
| Anisole | 100* |
| Benzothiazole | 62.4 |
| Chloroform | 26.0 |
| Ethyl Benzene | 100* |
| Ethyl Cinnamate | 100* |
| Leaded Gasoline | 100* |
| Methoxychlor | 84.2 |
| Methylene Chloride | 8.0 |
| Tributyl Phosphate | 51.2 |
| Unleaded Gasoline | 100* |
| Clofibric Acid | 0 |
| Coliphage (JJ resistant) | 100* |
| Coliphage (MS-2) | 100* |
| Attenuated Polio Virus | 100* |
| Latex Spheres (3μ) | 100* |

* none detected in process product water

## Activated Carbon Process Contaminant Removal Assessment

The primary function of the activated carbon process was to remove organic chemicals. Performance was tracked using total organic carbon as a measure of removal of these compounds. The broad spectrum of test methods (Table 11-3) used for organic compound analysis did detect several chemicals in this category. Most organic compounds were detected sporadically and at very low concentrations (<1 μg/L). A few compounds were detected frequently enough so that removal percentage could be calculated (Table 10-3).

Table 10-5 contains results for both first and second-stage activated carbon during various phases of plant operation. The percent removal listed is an average of mean values from all process configurations. Also, listed in the table are the results from the Contaminant Challenge Study (Chapter 7).

High removals (>75%) of many of the detected compounds illustrate the effectiveness of this process. Several of the detected chemicals were removed to undetectable levels. Three compounds (chloroform, bromoform, and tetrachloroethane) were found to desorb (discussed in Chapter 7) from the carbon columns at repeatable intervals of 7 or 11 weeks of operation. These substances were completely removed by membrane and air stripping treatment and, thus, were not detected in any plant product water sample.

**Table 10-3**
**Activated Carbon Process Organic Contaminant Removals**
% Removal
(Average of Mean Values)

| Contaminant | % Removal |
|---|---|
| Total Organic Carbon | 63.6 |
| Chloroform | 57.9* |
| 1,1,1-Trichloroethane | 89.0 |
| Trichloroethylene | 80.2 |
| Tetrachloroethene | 98.5 |
| Bromoform | 76.5* |
| Carbon Tetrachloride | 98.0 |
| Dichlorobromomethane | 100 |
| Dibromochloromethane | 100 |
| Tetrachloroethane | 42.9* |
| p-Dichlorobenzene | 100 |
| o-Dichlorobenzene | 84.9 |
| *Contaminant Challenge Study Results* | |
| Benzothiazole | 100 |
| Chloroform | 99.0 |
| Clofibric Acid | 100 |
| Methoxychlor | 100 |
| Methylene Chloride | 100 |
| Tributyl Phosphate | 100 |

*desorption occurred on an 11 week cycle for first-stage carbon and week 7 for second-stage

Two of the three organic chemicals that had not been effectively removed by lime clarification (clofibric acid and methylene chloride) were completely removed by activated carbon. Chloroform was the only compound that could be detected consistently following activated carbon treatment. The concentration remaining was near 1 μg/L and the substance was completely removed by the following membrane and air stripping processes. Activated carbon proved to be an effective barrier for many organic compounds.

## Reverse Osmosis and Ultrafiltration Contaminant Removal Assessment

Membrane processes were an essential part of the potable reuse treatment system. Reverse osmosis was included in the original Reuse Plant design and ultrafiltration was added (pilot-size system) before the start of the animal health effects study. The purpose of the ultrafiltration treatment sequence was to evaluate it as a possible less costly, 50/50 split treatment alternative to reverse osmosis that would produce water quality closer to Denver's current drinking water. A membrane was required for the split treatment to provide a barrier to microscopic organisms.

The removal capability of reverse osmosis is well known and confirmed by the Reuse Project test results some of which are listed in Table 10-4. This process has the ability to remove almost all classes of contaminants and is the only process included in the reuse treatment system to remove dissolved salts (e.g. sodium). Results for all parameter classes (general, inorganic, organic, microbiological, and radiological) are listed in Table 10-4. The results shown are mean values from samples taken during reuse treatment plant operation. Thus, removal could only be calculated for those substances present in the feed water.

This information was consequently augmented with the results from special studies. The Contaminant Challenge Study (Chapter 7) added high concentrations of a broad variety of compounds to evaluate the plant removal capability. The reverse osmosis (RO) system was evaluated for the removal of these compounds but again could only be evaluated for the removal of compounds that were detectable in the process feed and had, thus, survived the previous treatment processes. The ultrafiltration process was not yet in operation during this study.

A special study designed to evaluate the removal capability of the ultrafiltration (UF) system was also conducted later. The UF feed and product was tested for a long list of compounds. These results are listed in Table 10-4 under the heading *Ultrafiltration Special Study*.

Total dissolved solids results (General Parameters) from Reuse Plant operation during the animal health effects study showed that RO removed 96.8% while UF removed 34.7%. The RO results were expected while the UF result was surprisingly high. The UF special study revealed a high removal percentage for particle count 4-8 μm and color. Only minor removals for specific conductance, hardness, and alkalinity confirmed that the process was not designed to remove dissolved salts and minerals.

## Table 10-4
## Selected Contaminant Removals for RO and UF
% Removal of Contaminants Detected in Feed to Each Process
(Average of Geometric Mean Values)

| Contaminant | RO | UF | Contaminant | RO | UF |
|---|---|---|---|---|---|
| GENERAL | | | *Ultrafiltration Special Study Test Results* | | |
| Total Dissolved Solids | 96.8 | 34.7 | Ammonia | nt | 18.0 |
| *Ultrafiltration Special Study Test Results* | | | Sulfate | nt | 29.3 |
| Particle Count 4-8 μm (count/50 mL) | nt | 91.8 | Boron | nt | 5.3 |
| Specific Conductance ((μmhos) | nt | 11.6 | Nitrate-N | nt | 48.3 |
| Total Hardness | nt | 11.9 | Chloride | nt | 0 |
| Total Alkalinity | nt | 5.6 | Copper | nt | 0 |
| Color | nt | 100 | Fluoride | nt | 0 |
| INORGANIC | | | Iron | nt | 0 |
| Aluminum | 37.5 | 25.0 | Lithium | nt | 0 |
| Boron | 35.0 | 15.0 | Molybdenum | nt | 25 |
| Calcium | 98.5 | 38.0 | Nickel | nt | 50 |
| Chloride | 84.6 | 9.6 | Nitrate-N | nt | 48.3 |
| Copper | 18 | 9.0 | Orthophosphate-P | nt | 100 |
| Chromium | 100 | 100 | Total Phosphorus | nt | 25 |
| Fluoride | 100 | 20.9 | Silica | nt | 0 |
| Iron | 100 | 78.6 | Strontium | nt | 0 |
| Lithium | 100 | 31.2 | ORGANIC | | |
| Magnesium | 100 | 36.0 | Total Organic Halogen | 83.2 | 29.6 |
| Molybdenum | 100 | 75.0 | 2,2-dimethyl decane | 100 | 63.8 |
| TKN | 86.0 | 38.1 | Methyl-ethyl Propanoic acid | 100 | 100 |
| Ammonia-N | 83.7 | 27.6 | Tert-butylbenzene | 100 | 100 |
| Total Phosphate | 81.0 | 49.0 | *Contaminant Challenge Study Results* | | |
| Potassium | 94.8 | 27.6 | Chloroform | 40.0 | nt |
| Silica | 84.6 | 19.4 | MICROBIOLOGICAL | | |
| Sodium | 96.8 | 28.2 | *Contaminant Challenge Study Results* | | |
| Strontium | 100 | 36.8 | Coliphage (JJ) | 100 | nt |
| Sulfate | 98.9 | 61.1 | Coliphage (MS-2) | 100 | nt |
| *Contaminant Challenge Study Results* | | | Latex Spheres (3μ) | 100 | nt |
| Uranium | 100 | nt | *Ultrafiltration Special Study Test Results* | | |
| Chromium | 100 | nt | Coliphage (E. coli. 137) | nt | 100 |
| Arsenic | 100 | nt | Total Coliform | nt | 100 |
| | | | Nematodes | nt | 100 |
| Key | | | Giardia | nt | 100 |
| | | | RADIOLOGICAL | | |
| nt = not tested | | | Gross Beta Activity | 100 | 30.9 |
| | | | *Contaminant Challenge Study Results* | | |
| | | | Uranium | 100 | nt |
| | | | *Ultrafiltration Special Study Test Results* | | |
| | | | Gross Alpha | nt | 100 |
| | | | Gross Beta | nt | 16.7 |

All inorganic parameters were removed effectively by RO. The removal of these substances by UF was highly variable. Several parameters were not removed significantly (e.g. copper, chloride, aluminum, and boron) while others were successfully removed (iron, chromium, sulfate, nitrite, and nitrate).

Organic compounds found in the process feed water were removed by both RO and UF. Chloroform was an exception being removed by RO by only 40%. The air stripping process following both membrane processes, though, removed the small amount that remained.

All the microbiological organisms introduced into the reuse treatment plant and present in small amounts in the process feed water were removed completely by both RO and UF. UF was found to be a complete barrier to both nematodes and protozoan.

Only very low amounts of radioactive substances were present in the RO and UF feed water even during the special studies. Removals were difficult to calculate with confidence due to test method precision at these levels. The Gross Beta test showed lower removal for UF in both ambient and special study samples. Values present were well below any regulatory standard.

The test results from reuse treatment plant samples collected over more than two years of operation demonstrated the extensive removal capability of reverse osmosis. Dissolved salts, metals, organic compounds, microbiological organisms, and radiological substances were removed successfully by reverse osmosis. The ultrafiltration process removed far more contaminants than expected. Total dissolved solids and hardness were reduced significantly by this process. Many organic compounds were similarly removed. This membrane demonstrated in could serve as a companion treatment process to reverse osmosis and could form a barrier to both nematodes and protozoan.

## Multiple Barriers of Contaminant Removal

The reuse treatment plant design (Chapter 3) used a multiple barrier approach for ultimate safety and maximum contaminant removal. The individual and collective water quality test results confirm the superior capability of the Reuse Plant to remove all possible contaminants. Each treatment process was chosen primarily for its removal of certain pollutants. But extensive testing revealed that many treatment processes could also remove a variety of other substances and organisms.

Table 10-5 illustrates the contaminant removal barriers contained in the reuse treatment plant. Ultrafiltration and reverse osmosis are listed separately to show the relative removal capability of each. To obtain the total number of contaminant barriers only one is counted since they are not complimentary in the reuse treatment sequence. Every contaminant group has at least two barriers each of which can remove all the parameters in that category.

Each treatment process is a barrier for at least one contaminant group. Both lime clarification and reverse osmosis are barriers for all the contaminants. Ultrafiltration and ozone are effective for the removal of more than one contaminant group. This assessment illustrates the robust capability of the reuse treatment process to remove all possible contaminants of possible health concern.

**Table 10-5**
**Reuse Treatment Process Contaminant Removal Barriers**

| PROCESS BARRIER | CONTAMINANT GROUP | | | | |
|---|---|---|---|---|---|
| Treatment Process | Virus and Bacteria | Protozoa | Metals, Inorganics, and Radioactive | Organics | Process Barrier Total |
| Lime Clarification | + | + | + | + | 4 |
| Filtration | | + | | | 1 |
| UV | + | | | | 1 |
| Activated Carbon | | | ½ | + | 1½ |
| Reverse Osmosis | + | + | + | + | 4 |
| Ultrafiltration | + | + | ½ | ½ | 3 |
| Air Stripping | | | | ½ | 1 |
| Ozone | + | + | | | 2 |
| Chloramine | ½ | | | | 1 |
| **Total Contaminant Barriers** | 4½ | 4 | 2½ or 2 | 3½ | |

½ = partial barrier, removes a portion of the contaminant group

# 11
# Water Quality
# During Health Effects Study Period

The potable Reuse Demonstration Plant product water quality was compared to the existing high-quality drinking water enjoyed by Denver and to established national (Table 11-1) and international criteria. More than 10,000 samples were examined for chemical, physical, and microbiological contaminants to establish the water quality of the reuse treatment plant product water. The data tables (11-4, -5, -6, -7, -8) contain water quality test results from the health effects Reuse Plant treatment sequence (Figure 8-1) alongside those from Denver drinking water for the same time period. The calculated water quality values for a 50/50 blend of RO and UF treated reuse water is also presented for comparison with Denver drinking water.

Denver drinking water was chosen as the standard of comparison for two reasons: (1) Public acceptance would be enhanced if the potable reuse water were as good as or better than the existing acceptable potable supply; and (2) since Denver's drinking water will likely meet any future regulatory requirements this would insulate the Project from the uncertainties of the changing regulations defining safe drinking water.

## Table 11-1
## US EPA Drinking Water Regulations

(TT Treatment technic for removal but goal is zero. Turbidity is 0.3 NTU for 95% of monthly samples from combined filter effluent. Total Coliform is detected in 5% of all samples monthly and none are *E. coli* positive.)

| Contaminant | US EPA MCL (mg/L) | Contaminant | US EPA MCL (mg/L) |
|---|---|---|---|
| Microorganisms | | o-Dichlorobenzene | 0.6 |
| *Cryptosporidium* | TT | p-Dichlorobenzene | 0.075 |
| *Giardia lamblia* | TT | 1,2-Dichloroethane | 0.005 |
| *Legionella* | TT | 1,1-Dichloroethylene | 0.007 |
| Total Coliforms (including fecal coliform and *E. Coli*) | 5%/mo | cis-1,2-Dichloroethylene | 0.07 |
| Turbidity | TT (0.3) | trans-1,2-Dichloroethylene | 0.1 |
| Viruses (enteric) | TT | Dichloromethane | 0.005 |
| Disinfection By-Products | | 1,2-Dichloropropane | 0.005 |
| Bromate | 0.010 | Di(2-ethylhexyl) adipate | 0.4 |
| Chlorite | 1 | Di(2-ethylhexyl) phthalate | 0.006 |
| Haloacetic acids (HAA5) | 0.060 | Dinoseb | 0.007 |
| Total Trihalomethanes (TTHMs) | 0.080 | Dioxin (2,3,7,8-TCDD) | 0.00000003 |
| Disinfectants | | Diquat | 0.02 |
| Chloramines (as $Cl_2$) | 4.0 | Endothall | 0.1 |
| Chlorine (as $Cl_2$) | 4.0 | Endrin | 0.002 |
| Chlorine dioxide (as $ClO_2$) | 0.8 | Epichlorohydrin | TT |
| Inorganic Chemicals | | Ethylbenzene | 0.7 |
| Antimony | 0.006 | Ethylene dibromide | 0.00005 |
| Arsenic | 0.010 | Glyphosate | 0.7 |
| Asbestos (fiber > 10 micrometers) | 7 MFL | Heptachlor | 0.0004 |
| Barium | 2 | Heptachlor epoxide | 0.0002 |
| Beryllium | 0.004 | Hexachlorobenzene | 0.001 |
| Cadmium | 0.005 | Hexachlorocyclopentadiene | 0.05 |
| Chromium (total) | 0.1 | Lindane | 0.0002 |
| Copper | TT; Action Level=1.3 | Methoxychlor | 0.04 |
| Cyanide (as free cyanide) | 0.2 | Oxamyl (Vydate) | 0.2 |
| Fluoride | 4.0 | Polychlorinated biphenyls (PCBs) | 0.0005 |
| Lead | TT; Action Level=0.015 | Pentachlorophenol | 0.001 |
| Mercury (inorganic) | 0.002 | Picloram | 0.5 |
| Nitrate (measured as Nitrogen) | 10 | Simazine | 0.004 |
| Nitrite (measured as Nitrogen) | 1 | Styrene | 0.1 |
| Selenium | 0.05 | Tetrachloroethylene | 0.005 |
| Thallium | 0.002 | Toluene | 1 |
| Organic Chemicals | | Toxaphene | 0.003 |
| Acrylamide | TT | 2,4,5-TP (Silvex) | 0.05 |
| Alachlor | 0.002 | 1,2,4-Trichlorobenzene | 0.07 |
| Atrazine | 0.003 | 1,1,1-Trichloroethane | 0.2 |
| Benzene | 0.005 | 1,1,2-Trichloroethane | 0.005 |
| Benzo(a)pyrene (PAHs) | 0.0002 | Trichloroethylene | 0.005 |
| Carbofuran | 0.04 | Vinyl chloride | 0.002 |
| Carbon tetrachloride | 0.005 | Xylenes (total) | 10 |
| Chlordane | 0.002 | Radionuclides | |
| Chlorobenzene | 0.1 | Alpha particles | 15 picocuries per Liter (pCi/L) |
| 2,4-D | 0.07 | Beta particles and photon emitters | 4 millirems per year (about 50 pCi/L) |
| Dalapon | 0.2 | Radium 226 and Radium 228 (combined) | 5 pCi/L |
| 1,2-Dibromo-3-chloropropane | 0.0002 | Uranium | 0.030 |

## Water Quality Testing Program

For more than five years the Reuse Demonstration Plant was operated continuously to evaluate the treatment processes needed to produce potable water from unchlorinated secondary treated wastewater. Many treatment process alternatives were evaluated during the first half of this period. These evaluations provided the data needed to select the best treatment system that was then used during the two-year chronic toxicity and reproductive whole-animal health effects studies. Water quality test results were of primary importance for these evaluations and treatment process decisions.

During the two-year animal health effects study, water concentrates from the reuse treatment plant were continuously prepared and supplied to the animal health effects testing laboratory. The treatment process sequence could not be altered during this period or the results would be confounded. A comprehensive water quality testing program was conducted for this entire time that focused on the Reuse Plant influent and product water. This data was then complemented by the animal health effects study results to establish the safety of the reuse product water.

The water quality testing program included every known water contaminant (Table 11-2). Routine sampling was conducted as illustrated in the sampling schedule shown in Figure 5-1. The primary sample locations for this program were the plant influent, activated carbon effluent, reverse osmosis treatment plant effluent, ultrafiltration treatment plant effluent, and Denver drinking water. The activated carbon sample location was important since this was the feed water to both the reverse osmosis system and the ultrafiltration system. Process contaminant removals for these membrane treatment processes could, thus, be determined using this data. Numerous additional tests were conducted to support plant operations and for special evaluations.

Most chemical and physical test methods were found in the reference; *Standard Methods for the Examination of Water and Wastewater.* These were augmented with available methods for rare earth elements (spark source spectroscopy), and trace organic compounds. The methods used for trace organic compounds are listed in Table 11-3. These represent the known "best methods" at the time that could achieve the lowest detection levels and identify the broadest spectrum of possible compounds. These methods were suggested by the Project Expert Advisory Committee and supported by the US EPA research laboratory in Cincinnati, OH. US EPA test method numbers are listed in Table 11-3 but some of these methods were not yet approved at the time of the Reuse Demonstration Project.

Most water testing was conducted at the Department's Water Quality Control Laboratory. This facility had been augmented with the additional equipment and trained personnel to conduct the necessary testing to support the Reuse Project. The laboratory included a complete trace organic testing wing that contained GC/MS instrumentation and computerized data analysis system, several specialized gas chromatographs, volatile organic analyzers, and associated facilities for ultra-trace organic analysis. Also, a complete inorganic chemistry wing of the laboratory conducted atomic absorption spectroscopy for trace metal analysis and included a comprehensive wet chemistry testing capability. The laboratory had two separate wings for microbiological organism testing. The bacterial indicators, bacterial pathogens, and microscopic examinations were conducted in one laboratory while enteric viruses and coliphage were tested in the other. The bacteria laboratory was fully certified for testing drinking water compliance samples.

The virus testing facility had begun development in 1975, even before the Project was officially started. Training was conducted at the Center for Disease Control and Prevention in Atlanta and the US EPA laboratories in Cincinnati. Clean rooms and biohazard hoods were incorporated into this facility. Before beginning analysis of reuse samples the virus laboratory split samples with US EPA and several contract laboratories to verify its ability to obtain reliable results. While active, the Denver Water Department was the only water supplier in the country with an enteric virus testing laboratory.

To augment this extensive in-house water quality testing capability, contract laboratories were utilized for some of the specialized and infrequently performed tests. These included *Giardia and cryptosporidium,* radioactive isotopes, rare earth elements, haloacetic acids, and aldehydes. The contract laboratories provided extensive quality assurance data along with the test results.

## Table 11-2
## Water Quality Analytical Testing Program
## Substances Routinely Analyzed
(Trace Organic Test Methods Table 11-3)

| GENERAL | Asbestos | Enteric Virus |
|---|---|---|
| Total Alkalinity | Zinc | *Entamoeba histolytica* |
| Total Hardness | Sodium | *Cryptosporidium* |
| TSS | Lithium | Algae |
| TDS | Titanium | *Clostridium perf.* |
| Specific Conductance | Barium | *Shigella* |
| pH | Silver | *Salmonella* |
| Turbidity | Rubidium | *Campylobacter* |
| Particle Count | Vanadium | *Legionella* |
| Temperature | Iodide | ORGANIC |
| Dissolved Oxygen | Antimony | Total Organic Carbon |
| Color | Beryllium | Total Organic Halogen |
| Odor | Iridium | MBAS |
| INORGANIC | Cobalt | Trihalomethanes |
| Aluminum | Thorium | Haloacetic acids |
| Arsenic | Tellurium | Aldehydes |
| Boron | Bismuth | Methylene Chloride |
| Bromide | Niobium | Tetrachloroethene |
| Cadmium | Tin | 1,1,1-Trichloroethane |
| Calcium | Osmium | Trichloroethene |
| Chloride | Tungsten | 1,4- Dichlorobenzene |
| Chromium | Cesium | Chloroform |
| Copper | Palladium | RADIOLOGICAL |
| Cyanide | Platinum | Gross Alpha |
| Fluoride | Zirconium | Gross Beta |
| Iron | Rhodium | Radium 228 |
| Potassium | Gallium | Radium 226 |
| Magnesium | Germanium | Tritium |
| Manganese | Ruthenium | Radon 222 |
| Mercury | Gold | Plutonium-total |
| Molybdenum | MICROBIOLOGICAL | Uranium-total |
| TKN | m-HPC | RARE EARTH ELEMENTS |
| Ammonia-N | Total Coliform | Lanthanum, Terbium |
| Nitrate-N | Fecal Coliform | Cerium, Dysprosium |
| Nitrite-N | Fecal Strep | Praseodymium, Holmium |
| Nickel | Coliphage B | Neodymium. Erbium |
| Total Phosphorous | Coliphage C | Promethium, Thulium |
| Selenium | Giardia | Samarium, Ytterbium |
| Silica | Endamoeba coli | Europium, Lutetium |
| Strontium | Nematodes | Gadolinium, Actinides |
| Sulfate | | |
| Lead | | |

**Table 11-3**
**Test Methods Used for Broad Spectrum Trace Organic Analyses**

| Test Method | Testing Frequency Comments |
|---|---|
| Volatile Organics (EPA 502.2) | Primary test methods: influent sampled every 18 days for more than five years, treatment plant effluents sampled every 6 days, Denver water sampled every 18 days for more than five years |
| Grob Closed Loop Stripping- GC/MS (EPA 8270) | |
| Pesticides (EPA 508) | These test methods were used on all samples quarterly. Additional tests were performed when indicated by results from the test methods above. |
| Herbicides (EPA 515.1) | |
| Carbamate Pesticides (EPA 531.1) | |
| Polychlorinated Biphenyls (EPA 505) | |
| Polynuclear Aromatic Hydrocarbons (EPA 610) | |
| Base, Neutral & Acid Extractables (EPA625) | |
| Trihalomethanes (EPA 501.1) | |
| Haloacetic Acids (EPA 552) | |
| Aldehydes (EPA 556.1) | |
| Disinfection byproducts (EPA 551.1) | |

# Water Quality Test Results

The water quality test results are divided into five groups to facilitate the evaluation: general parameters, inorganic parameters, organic parameters, microbiological organisms, and radiological parameters. The test result tables display geometric mean values for all routinely tested parameters. The symbol "<" is used to indicate that the value is below the method detection limit and the value shown after the symbol is the detection limit. The values in the 50/50 Blend RO and UF Plant Effluent column are calculated from the values for RO and UF. Calculations for a few of the parameters required some estimation due to the units used but although these may depart slightly from actual values they are adequately accurate for the rough comparisons used in this discussion.

### General Parameters

This group contains physical and aggregate chemical measures of water quality (Table 11-4). Several of these are familiar to potential customers and often define water quality to the public. Included in this analysis group are: pH, hardness, alkalinity, specific conductance, temperature, dissolved oxygen, total dissolved solids, total suspended solids, color, particle count, odor, and turbidity.

As noted in the plant operation discussion (Chapter 8), the ultrafiltration pilot-size system was uncharacteristically impacted by the ozone gas delivery system and the inefficiency of small chlorine disinfection contactor. Turbidity and particles were added to the ultrafiltration product water by these systems but this would not be the case for a full-scale system. Even with these inefficiencies the 4-8 $\mu$m particle count for the blended product would be less than Denver drinking water. Turbidity too was still low and this value (0.13 NTU) would be substantially lower (probably less than 0.06 NTU like reverse osmosis) for a larger scale system.

Temperature of the Reuse Plant product was warmer than Denver drinking water on average. But the difference varied seasonally. Denver drinking water temperature rises in the summer reaching levels nearly equal to the Reuse Plant effluent but in the winter the difference was far greater. This variance would be noticeable to customers but blending in the drinking water distribution network would mitigate this difference.

The mineral content of RO treated reuse water would be very low by any comparison and would likely require chemical addition to reduce corrosion potential. The UF treated reuse water would be higher in mineral content (measured by hardness, alkalinity, and specific conductance) than Denver drinking water. The UF treated water might still be acceptable to customers, and would probably be non-corrosive. Ultrafiltration lowered hardness by 38% and even total dissolved solids by 35%. The 50/50 blended RO and UF treated water would be somewhat lower in mineral content than Denver drinking water. It was expected that this would be viewed favorably by most consumers.

**Table 11-4**
**Water Quality Results**
**Jan 9, 1989-Dec 20, 1990**
**General Parameters**
(geometric mean values in mg/L unless otherwise noted)

| Parameter | Plant Influent | Plant Effluent RO process sequence | Plant Effluent UF process sequence | Denver Drinking Water | 50/50 Blend RO and UF Plant Effluent |
|---|---|---|---|---|---|
| Total Alkalinity | 249 | 2 | 154 | 64 | 78 |
| Total Hardness | 206 | 4 | 101 | 107 | 53 |
| Total Suspended Solids | 12 | <1 | <1 | <1 | <1 |
| Total Dissolved Solids | 581 | 17 | 342 | 183 | 180 |
| Specific Conductance (μmhos/cm) | 983 | 60 | 661 | 294 | 361 |
| pH (units) | 6.8 | 6.6 | 7.7 | 7.8 | 7.2 |
| Dissolved Oxygen | 3.5 | 8.4 | 7.1 | nt | 7.8 |
| Temperature (°C) | 20 | 22 | 22 | 6 | 22 |
| Turbidity (NTU) | 9.2 | 0.06 | 0.2 | 0.3 | 0.13 |
| Color (units) | 9 | <1 | <1 | <1 | <1 |
| Particle Count >128 μm (count/50 mL) | nt | <1 | <1 | <1 | <1 |
| Particle Count 64-128 μm (count/50 mL) | nt | 1.5 | 2.6 | 1.6 | 2.0 |
| Particle Count 32-64 μm (count/50 mL) | nt | 25 | 55 | 41 | 40 |
| Particle Count 16-32 μm (count/50 mL) | nt | 78 | 230 | 233 | 154 |
| Particle Count 8-16 μm (count/50 mL) | nt | 163 | 903 | 869 | 533 |
| Particle Count 4-8 μm (count/50 mL) | nt | 252 | 2286 | 2274 | 1269 |
| Odor (TON) | >200 | <1 | <1 | <2 | <1 |

The results of general parameters analysis show very low values for water treatment through the RO process sequence. The UF process sequence yielded higher values but the 50/50 blend of the RO and UF products compared most favorably to Denver drinking water. Water hardness for the blend would still be about 50% less than Denver's water but most customers would likely react to this difference positively.

### Inorganic Parameters

Metals, cations, and anions make up the inorganic parameter category. Of the more than seventy substances included in this category none were found at levels approaching regulatory standards (Table 11-1) in any sample including the untreated plant influent.

The only metals detected in the RO treatment sequence product water were aluminum, boron, calcium, potassium, sodium, and zinc. All these elements are unregulated and the concentrations were well below any level of possible concern. The UF treatment sequence product also didn't contain any metals at concentrations nearing any health standard. Calcium and sodium were both higher in the UF product than Denver drinking water. The calcium level of the 50/50 blend is comparable to Denver drinking water but the sodium level is somewhat higher (42 mg/L vs. 19 mg/L) while still far below any level of concern.

Anions found in the reuse products included: chloride, fluoride, and phosphate. Only reverse osmosis reduced the chloride level substantially. The concentration found in the UF product and calculated for the 50/50 blend were both far below the US EPA secondary drinking water standard of 250 mg/L. Fluoride is naturally occurring in part of Denver's water supply and it is augmented to maintain 0.7 mg/L in the rest of the system. The amount found in the UF product is below the Denver drinking water level as is the blend.

Ammonia Results Discussion

As shown in Table 11-5, the only inorganic substance significantly higher in reuse treatment plant product water than Denver drinking water was ammonia-nitrogen. This difference was not of any consequence for the Demonstration Project because the animal health effects study focused on the organic substances in the water. To provide information for the design of a full-scale reuse treatment plant, additional methods of nitrogen removal were examined.

Nitrogen is potentially present in three forms: nitrite ($NO_2^-$), nitrate ($NO_3^-$), and ammonia ($NH_3$ or $NH_4^+$). Both nitrite and nitrate are regulated and have known health consequences. The Maximum Contaminant Level (MCL) for nitrate-N is 10 mg/L and the MCL for nitrite-N is 1.0 mg/L. Ammonia was the predominant nitrogen form in conventional secondary wastewater effluent and is unregulated by the US EP A. The European communities (EU) have established a Maximum Admissible Concentration (MAC) for ammonia at 0.5 mg/L while potable reuse projects other than the Denver Project have set ammonia-nitrogen limits which range from 0.5 mg/L to 4 mg/L.

The primary ammonia barrier incorporated into the Reuse Plant design was selective ion exchange utilizing the naturally occurring zeolite, *clinoptilolite*. During start-up operations it was determined that the system was unable to achieve the design goal of less than 1 mg/L ammonia-nitrogen. In addition, the operational cost for the ion exchange treatment was estimated at $0.58 per 1,000 gallons. The total cost including amortized capital was estimated at between $0.93 and $1.20 per 1,000 gallons. This represented a sizeable cost for the removal of a single contaminant that is not a health concern.

Another method of removing ammonia needed to be identified to prepare estimates of cost for a proposed full-scale potable reuse treatment plant. An extensive literature review and subsequent laboratory tests identified biological nitrification-denitrification as a promising treatment process. Ammonia can be transformed, using this process, by aerobic bacteria to nitrite and under favorable conditions then sequentially converted to nitrate. These nitrogen forms can then be converted to nitrogen gas by anaerobic bacteria. The nitrogen gas then escapes from the water into the air.

The biological nitrogen removal processes, thus, was investigated under a research grant at the University of Colorado Department of Environmental Engineering. The resulting design, tested at pilot scale, involved the fixed-film nitrification and denitrification using acetate (acetic acid) as the organic carbon source. The pilot system was able to achieve superior nitrogen removal compared to the ion exchange system at a fraction of the cost.

Operation of the Reuse Demonstration Plant without nitrification and denitrification did not compromise the goals of the Project since the health effects of ammonia were well-known and minimal. Nevertheless, a nitrified and partially denitrified secondary effluent was assumed as the eventual raw water source from the wastewater treatment plant. In any case, an alternative treatment option was demonstrated which could be utilized to remove nitrogen species if it were necessary. This process was included in the cost estimates for a future full-scale potable water reuse treatment plant.[3]

---

[3] As of the publication of this report the ammonia levels in the wastewater treatment plant effluent have already been reduced by more than 50% and the plant was planning on meeting future regulatory limits that will require ammonia reduction, as predicted, by more than 90%.

## Table 11-5
## Water Quality Results
## Jan 9, 1989-Dec 20, 1990
## Inorganic Substances
(geometric mean values in mg/L unless otherwise noted)

| Parameter | Plant Influent | Plant Effluent RO process sequence | Plant Effluent UF process sequence | Denver Drinking Water | 50/50 Blend RO and UF Plant Effluent |
|---|---|---|---|---|---|
| Aluminum | 0.057 | 0.010 | 0.012 | 0.144 | 0.011 |
| Arsenic | <0.001 | <0.001 | <0.001 | <0.001 | <0.001 |
| Boron | 0.41 | 0.23 | 0.34 | 0.13 | 0.29 |
| Bromide | <0.1 | <0.1 | <0.1 | <0.1 | <0.1 |
| Cadmium | <0.001 | <0.001 | <0.001 | <0.001 | <0.001 |
| Calcium | 52.1 | 0.8 | 32.6 | 25.9 | 16.7 |
| Chloride | 97 | 16 | 94 | 25 | 55 |
| Chromium | 0.002 | <0.001 | <0.001 | <0.001 | <0.001 |
| Copper | 0.023 | 0.009 | 0.010 | 0.005 | 0.009 |
| Cyanide | <0.01 | <0.01 | <0.01 | <0.01 | <0.01 |
| Fluoride | 1.2 | <0.1 | 0.6 | 0.7 | 0.3 |
| Iron | 0.25 | <0.001 | 0.068 | 0.028 | 0.034 |
| Potassium | 12.7 | 0.6 | 8.4 | 2.0 | 4.5 |
| Magnesium | 12.6 | <0.2 | 1.6 | 7.9 | 0.9 |
| Manganese | 0.10 | <0.01 | <0.01 | 0.01 | <0.01 |
| Mercury | 0.0001 | <0.00005 | 0.0001 | 0.0001 | 0.00005 |
| Molybdenum | 0.021 | <0.002 | 0.005 | 0.012 | 0.002 |
| TKN | 26.6 | 3.7 | 16.4 | 0.8 | 10.1 |
| Ammonia-N | 24.6 | 3.9 | 17.3 | 0.6 | 10.6 |
| Nitrate-N | 0.1 | 0.1 | 0.1 | <0.1 | 0.1 |
| Nitrite-N | <0.1 | <0.1 | <0.1 | <0.1 | <0.1 |
| Nickel | 0.007 | <0.001 | <0.001 | <0.001 | <0.001 |
| Total Phosphate | 5.4 | 0.02 | 0.05 | 0.01 | 0.03 |
| Selenium | <0.001 | <0.001 | <0.001 | <0.001 | <0.001 |
| Silica | 13.6 | 1.7 | 8.7 | 6.1 | 5.2 |
| Strontium | 0.39 | <0.01 | 0.12 | 0.23 | 0.06 |
| Sulfate | 166 | 2 | 58 | 47 | 30 |
| Lead | <0.001 | <0.001 | <0.001 | <0.001 | <0.001 |
| Uranium | 0.004 | <0.0006 | <0.0006 | 0.002 | <0.0006 |
| Zinc | 0.38 | 0.005 | 0.11 | 0.003 | 0.058 |
| Sodium | 117 | 4 | 79 | 19 | 42 |
| Lithium | 0.17 | <0.001 | 0.011 | 0.007 | 0.005 |
| Titanium | 0.05 | <0.01 | 0.02 | <0.01 | 0.01 |
| Barium | 0.03 | <0.01 | <0.01 | 0.03 | <0.01 |
| Silver | 0.001 | <0.001 | <0.001 | <0.001 | <0.001 |
| Rubidium | 0.002 | <0.001 | <0.001 | <0.001 | <0.001 |
| Vanadium | 0.002 | <0.001 | <0.001 | <0.001 | <0.001 |
| Iodide | <0.01 | <0.01 | <0.01 | <0.01 | <0.01 |
| Antimony | <0.001 | <0.001 | <0.001 | <0.001 | <0.001 |
| Asbestos (MFibers/L) | 1.1 | <0.1 | <0.1 | 0.1 | <0.1 |

**Table 11-5 continued**
Inorganic parameters tested but results were below detection limit and mean values could not be calculated.

| | |
|---|---|
| Beryllium | Hafnium |
| Iridium | Holmium |
| Cobalt | Terbium |
| Thorium | Lanthanum |
| Tellurium | Lutetium |
| Bismuth | Neodymium |
| Niobium | Thulium |
| Tin | Cerium |
| Osmium | Dysprosium |
| Tungsten | Yterbium |
| Cesium | Erbium |
| Palladium | Praseodymium |
| Platinum | Yttrium |
| Zirconium | Europium |
| Rhodium | Gadolinium |
| Gallium | Samarium |
| Germanium | Scandium |
| Ruthenium | Gold |

### *Organic Parameters*

Trace organic analysis methods were the subject of discussions with the expert advisory committee and the Department's consultants. The methods needed to be capable of detecting a broad spectrum of possible organic compounds at extremely low concentrations. The consensus of the experts was to use a suite of methods (Table 11-3) both *Standard* (included in the latest edition of *Standard Methods for the Examination of Water and Wastewater*) and emerging. Emerging methods were those that were proven but had not yet been accepted in *Standard Methods or US EPA methods*.

Two methods chosen for screening (reuse product waters sampled every 6 days) samples for trace organic compounds were the Grob closed loop stripping and volatile organic procedures. The Grob closed loop stripping method coupled with gas chromatography and mass spectrometry was an emerging method that had the ability to detect and identify a wide variety of organic compounds at sub-part-per-billion concentrations. The volatile organic procedure was a *Standard* method routinely used for this class of compounds. These compounds were expected to be somewhat resistant to reuse water treatment and included substances of health significance.

Samples from the Reuse Plant and Denver drinking water were tested according to the sampling schedule (shown in Figure 8-1). Twenty-four hour composite samples were used for most of the routine testing program. The sampling devices and quality control procedures were discussed in Chapters 5 and 7). Additional trace organic tests were conducted to support special studies like activate carbon removal of volatile organic compounds and to investigate process performance issues like those encountered with air stripping (Chapter 8). Over the two-year duration of the animal health effect study 1,487 routine organic test procedures were conducted to produce the results presented in Table 11-6.

The US EPA representatives on the expert advisory committee explained that new toxicological information was going to result in some additional compounds being considered for regulation. Based on this information several additional organic test methods were added to the list of routine analyses these included, haloacetic acids, aldehydes, and disinfection byproducts. Methods for the analysis of these compounds were developed at US EPA laboratories but had not been approved. Health standards were subsequently established for two classes of these compounds: trihalomethanes MCL = 80 µg/L and haloacetic acids MCL = 60 µg/L. All the methods of analysis later became approved US EPA methods (Table 11-3).

Total organic carbon was used to evaluate the reuse treatment system processes for the removal of organic compounds. Both the RO and UF treatment sequences produced product water lower in total organic carbon than Denver drinking water. The calculated value for a blended product was less than a third of the drinking water concentration. The potential for the formation of chlorinated disinfection byproducts, as a result, would likely be lower in the reuse product waters. Although this is probably a valid conclusion, since chloramines were used as the secondary disinfectant the formation of these potential carcinogens was avoided.

## Table 11-6
## Water Quality Results
## Jan 9, 1989-Dec 20, 1990
## Organic Substances
(geometric mean values in mg/L unless otherwise noted)

| Parameter | Plant Influent | Plant Effluent RO process sequence | Plant Effluent UF process sequence | Denver Drinking Water | 50/50 Blend RO and UF Plant Effluent |
|---|---|---|---|---|---|
| Total Organic Carbon | 16.3 | 0.2 | 1.1 | 2.1 | 0.6 |
| Total Organic Halogen | 0.109 | 0.006 | 0.024 | 0.046 | 0.015 |
| MBAS | 0.4 | <0.1 | <0.1 | <0.1 | <0.1 |
| Trihalomethanes | 0.0029 | <0.0005 | <0.0005 | 0.0039 | <0.0005 |
| Haloacetic acids | <0.001 | <0.001 | <0.001 | 0.0039 | <0.001 |
| Methylene Chloride | 0.0174 | <0.0005 | <0.0005 | <0.0005 | <0.0005 |
| Tetrachloroethene | 0.0096 | <0.0005 | <0.0005 | <0.0005 | <0.0005 |
| 1,1,1-Trichloroethane | 0.0027 | <0.0005 | <0.0005 | <0.0005 | <0.0005 |
| Trichloroethene | 0.0007 | <0.0005 | <0.0005 | <0.0005 | <0.0005 |
| 1,4- Dichlorobenzene | 0.0021 | <0.001 | <0.001 | <0.001 | <0.001 |
| Formaldehyde | <0.005 | <0.005 | 0.0124 | <0.005 | 0.006 |
| Acetaldehyde | 0.0095 | <0.005 | 0.0072 | <0.005 | <0.005 |
| Dichloroacetic acid | 0.001 | <0.0001 | <0.0001 | 0.0039 | <0.0001 |
| Trichloroacetic acid | 0.0056 | <0.0001 | <0.0001 | <0.0001 | <0.0001 |
| Chloroform | 0.0029 | <0.0005 | <0.0005 | 0.0029 | <0.0005 |
| Bromodichloromethane | <0.0010 | <0.0010 | <0.0010 | 0.0010 | <0.0010 |
| 1,1-Dichloropropanone | <0.0005 | <0.0005 | <0.0005 | 0.0006 | <0.0005 |

An aggregate measure of halogenated organic compounds is the total organic halogen test. Low levels were found in all product waters and Denver drinking water. Although there isn't a standard for this measurement, it is believed to be an indicator for individual halogenated compounds that require lengthy test procedures. Two regulated compounds classes in this group, trihalomethanes and haloacetic acids, were not detected in either (RO or UF) reuse treatment product water.

Ozone oxidation of organic compounds produces aldehydes. These compounds are not normally analyzed in drinking water but since the reuse process used ozone these became substances of interest. Both formaldehyde and acetaldehyde were found in the UF treatment sequence product water but the concentrations were very low. These compounds are not regulated and there are no known health consequences.

Denver water contains small concentrations of volatile organic chlorinated disinfection byproducts. These compounds are not present in amounts approaching health standards and they were not detected in the reuse treatment product water. They were important though for the animal health effects study. Volatile compounds are lost in the sample concentration procedure for the health effects study. These compounds thus were necessarily added back to the test samples in the concentrations found in Denver water (Chapter 12). The amounts added were based on test results from testing conducted for several years before starting the animal health effects study. No volatile organic compounds were detected in either of the reuse treatment product waters so no compounds needed to be added to these health effects samples.

The only organic compounds consistently detected in the reuse treatment product waters or Denver drinking water are shown in Table 11-6. The many trace organic test procedures listed in Table 11-3 detected additional organic compounds sporadically and at levels mostly below 1 µg/L. Of all the compounds detected only thirty were present in more than 5% of the plant influent samples. This number was reduced to only seventeen after activated carbon treatment and none were found at this level in either the RO or UF treatment sequence product waters. Only two compounds were detected in more than five percent of the Denver drinking water samples. Infrequently detected compounds with concentrations above 1 µg/L were absent from Denver drinking water and RO treatment sequence samples. Only two compounds at this level were detected in the UF treatment sequence samples and neither of these was regulated or had any known health consequences. After analyzing reuse treatment samples for a broad spectrum of possible organic substances over several years no compounds of concern were detected. The reuse product water compared favorably to Denver drinking water in regard to organic content.

### *Microbiological parameters*

The unchlorinated secondary wastewater used for the influent to the Reuse Plant contained high levels of microbiological contaminants. Nearly every pathogen tested was found in the Reuse Plant influent (Table 11-7). All microbial pathogens were completely eliminated by the high pH lime clarification process. Even when extremely high amounts of resistant coliphage, attenuated polio virus, and 3µm latex spheres were added to the Reuse Plant in the contaminant challenge studies, 100% were removed by the lime clarification treatment process.

Membrane heterotrophic plate count (mHPC), a general bacteria indicator, is the only microbiological measurement that gave positive results for a Reuse Plant product water and Denver drinking water. This group of bacteria is not necessarily pathogenic and there is no health standard for mHPC. As explained earlier (Chapter 8) the ultrafiltration pilot system had an inefficient disinfection module due to its small size. The mHPC levels found, even so, were well below those commonly encountered for many drinking water supplies.

**Table 11-7**
**Water Quality Results**
**Jan 9, 1989-Dec 20, 1990**
**Microbiological Organisms**
(geometric mean values in mg/L unless otherwise noted)

| Parameter | Plant Influent | Plant Effluent RO process sequence | Plant Effluent UF process sequence | Denver Drinking Water | 50/50 Blend RO and UF Plant Effluent |
|---|---|---|---|---|---|
| Total Coliform (count/100 mL) | $5.9 \times 10^5$ | <0.2 | <0.2 | <0.2 | <0.2 |
| m-HPC (count/mL) | $1.1 \times 10^6$ | <0.01 | 182 | 2.8 | 91 |
| Fecal Strep (count/100 mL) | $8.1 \times 10^3$ | <0.2 | <0.2 | <0.2 | <0.2 |
| Fecal Coliform (count/100 mL) | $6.2 \times 10^4$ | <0.2 | <0.2 | <0.2 | <0.2 |
| *Shigella* (present or absent) | present | absent | absent | absent | absent |
| *Salmonella* (present or absent) | present | absent | absent | absent | absent |
| *Clostridium perfringens* (count/100 mL) | $8.5 \times 10^3$ | <0.2 | <0.2 | <0.2 | <0.2 |
| *Campylobacter* (present or absent) | present | absent | absent | absent | absent |
| *Legionella* (present or absent) | present | absent | absent | absent | absent |
| *Giardia* (cysts/L) | 1.8 | <0.01 | <0.01 | <0.01 | <0.01 |
| *Cryptospordium* (oocysts/L) | 0.4 | <0.01 | <0.01 | <0.01 | <0.01 |
| *Entamoeba histolytica* (cysts/L) | <0.1 | <0.01 | <0.01 | <0.01 | <0.01 |
| *Endamoeba coli* (cysts/L) | 1.6 | <0.01 | <0.01 | <0.01 | <0.01 |
| Helminths (count/L) | <0.1 | <0.01 | <0.01 | <0.01 | <0.01 |
| Nematodes (count/L) | 4.1 | <0.01 | <0.01 | <0.01 | <0.01 |
| Algae (count/L) | 1.5 | <0.01 | <0.01 | <0.01 | <0.01 |
| Coliphage- B host (pfu/100 mL) | $2.1 \times 10^4$ | <0.1 | <0.1 | <0.1 | <0.1 |
| Coliphage- C host (pfu/100 mL) | $5.3 \times 10^4$ | <0.1 | <0.1 | <0.1 | <0.1 |
| Enteric Virus (MPNIU/10L) | Not tested | <0.01 | <0.01 | Not tested | <0.01 |

Viruses were of particular concern when the Demonstration Project began. To address this issue enteric virus testing capability was developed to support the Project. A separate laboratory was constructed containing clean rooms with biohazard hoods. An analyst was trained to perform the tests that used living cells and required clinical techniques. The test method was quite involved and required an enormous commitment of personnel. To take

124

maximum advantage of these resources the testing focused on the reuse treatment RO and UF product waters. Previous testing had confirmed the presence of enteric viruses (20-500 MPNIU/10L) in the plant influent so additional testing was unnecessary. No enteric virus was detected in any reuse treatment product water after more than five years of testing.

A bacterial virus, coliphage, served to provide additional information on the removal of these organisms. Two hosts were used to evaluate coliphage reaction to reuse treatment. Coliphage was found to be present in high concentrations in the Reuse Plant influent. These contaminants were not found in any Reuse Treatment Plant product water.

Microbiological organisms were present in extremely high levels in the Reuse Treatment Plant influent. High pH lime clarification completely removed these pathogens. The Reuse Plant treatment sequence contained several other processes (ozone, membrane treatment either reverse osmosis or ultrafiltration, and UV) that each could remove high levels of these organisms. Reuse treatment provided an extreme level of protection from microbiological organisms far exceeding disinfection processes included in conventional water treatment plants.

### Radiological Parameters

There are sources of radioactive substances in the Denver area. These include uranium, plutonium, and radon 222. This knowledge necessitated analysis of these substances along with other radioactive isotopes (Table 11-8). The Gross Alpha and Gross Beta activity tests were routinely performed to assess the presence and removal of radioactive substances. While the initial values were low they were further reduced by reuse treatment to levels comparable to those found in Denver drinking water. The values encountered were orders of magnitude below regulatory limits.

## Table 11-8
## Water Quality Results
## Jan 9, 1989-Dec 20, 1990
## Radiological Parameters
(geometric mean values in pCi/L)

| Parameter | Plant Influent | Plant Effluent RO process sequence | Plant Effluent UF process sequence | Denver Drinking Water | 50/50 Blend RO and UF Plant Effluent |
|---|---|---|---|---|---|
| Gross Alpha | 2.9 | <0.1 | <0.1 | 1.3 | <0.1 |
| Gross Beta | 10.0 | <0.4 | 5.6 | 2.3 | 2.8 |
| Radium 228 | <1 | <1 | <1 | <1 | <1 |
| Radium 226 | <0.3 | <0.3 | <0.3 | <0.3 | <0.3 |
| Tritium | <100 | <100 | <100 | <100 | <100 |
| Radon 222 | <20 | <20 | <20 | <20 | <20 |
| Plutonium-total | <0.02 | <0.02 | <0.02 | <0.02 | <0.02 |
| Uranium-total (mg/L) | 0.004 | <0.0006 | <0.0006 | 0.002 | <0.0006 |

## Conclusions

The Reuse Treatment Plant product waters processed either by reverse osmosis or ultrafiltration were examined for a multitude of pollutants. General parameters, radioactive substances, and inorganic chemicals results verified the reuse treatment capability to remove or eliminate any of these substances. Organic chemicals were evaluated using twelve separate analytical test procedures that included gas chromatography-mass spectrometry identification capability. These comprehensive evaluations again proved that all pollutants were completely removed or reduced to levels below any possible concern. The reuse treatment sequence was most impressive when evaluating microbiological contaminants. A comprehensive list of pathogens were tested and detected at high levels in the plant influent. None were detected in the Reuse Plant product waters. In fact none survived the first treatment process step, lime clarification, and at least three more pathogen barrier processes remained each of which could removal all microbiological contaminants.

The high quality of the direct potable reuse product water was demonstrated for a comprehensive list of contaminants. Hundreds of substances were analyzed over a period covering several years. More than 10,000 samples were taken to include hourly, daily, weekly, and seasonal variations in water quality. The results from all these tests confirm the ability of the reuse treatment to remove contaminants and to produce water satisfying all regulatory standards. Further, the product waters from the Reuse Treatment Plant meet or exceed the quality of Denver's current drinking water.

# 12

# Whole-Animal Health Effects Study

The need for a whole-animal health effects study was established early in the conceptual stage of the Reuse Demonstration Project. To achieve the goal of establishing the unquestioned safety of the reuse treatment plant product water it was recognized that this type of study would be required to augment the results of a comprehensive water quality testing program.

The Project was built on the successes of earlier reuse and advanced wastewater treatment projects that had conducted various health effects studies. The direct water reuse treatment plant in Windhoek, Namibia had been in operation since 1968. Extensive research had been conducted there including epidemiological studies to assess any adverse health effects on the population. Indirect potable reuse practiced at the Orange County Water District Water Factory 21 (1971) had carried out *in vitro* mutagenicity studies (using bacteria or cell transformation methods) that showed no adverse effects

The US EPA issued a policy statement on water reuse in 1972. This statement did not support direct potable reuse. However, reclaiming treated wastewater to augment drinking water supply reservoirs was allowed assuming research and test results had shown that the public health was not jeopardized. US EPA did state its support for continued research and demonstration projects "*including epidemiological and toxicological analyses of effects, advanced waste and drinking water treatment process design and operation, development of water quality requirements for various reuse opportunities, and cost-effectiveness studies.*" This statement formed the basis for US EPA support for the Denver Demonstration Project when it was proposed in 1979 and illustrated the agency's interest in toxicological health effects studies.

The earliest mention (in the literature) of the need for whole-animal health effects testing for the Denver Reuse Demonstration Project was in a 1976 article by Work, et al. This article followed the completion of the plant conceptual design and described a yet unfunded Project to demonstrate the feasibility of direct potable reuse. This article commented on the need for health effects testing: "*Health studies will still be required even with use-increment removal. These will serve as a back-up to quality testing and will consist primarily of toxicological studies performed on the demonstration plant effluent. Early efforts involving literature review and design of the health studies will begin concurrently with demonstration plant design. Actual studies will begin with operation of the plant and continue a minimum of five years or until safety is adequately demonstrated. Coordination with a national research program is planned."*

As the Denver Demonstration Project sought US EPA funding first in 1976 and then successfully in 1979 but before the health effects study started (1988), several other reuse projects evolved so that their approach to chronic health effects could be considered. Upper Occoquan Service Authority in Virginia (1978) started treating wastewater using advanced methods to serve as a supply for its potable water treatment plant. Several *in vitro* studies were part of the testing program and no adverse effects were found. The Potomac Estuary Experimental Water Treatment Plant (circa. 1980) used many treatment processes to evaluate the product for a possible drinking water source for Washington, D.C. This project used several mutagenicity methods to assess carcinogenic potential. Tampa, Florida conducted an extensive evaluation of advance wastewater treatment (circa. 1987) for possible use as a potable supply. This project conducted several types of mutagenicity studies and a sub-chronic (90 day) whole-animal health effects study using rats and mice.

The National Research Council of the National Academy of Science Panel on Quality Criteria for Water Reuse, Board on Toxicology and Environmental Health Hazards, Commission on Life Sciences (NRC) published *Quality Criteria for Water Reuse* in 1982. This document prepared by a panel of experts outlined the need for health effects bioassays to support projects where human consumption was being considered. The committee recommended a three-tier analysis process culminating in a chronic lifetime whole-animal health effect study. The first two tiers of testing were primarily for screening purposes used to determine toxicity and carcinogenicity potential before committing to the expense of a lifetime animal study. The panel concluded: *"The ultimate evaluation of the potential adverse health effects from reused water must come from chronic bioassays in whole animals."*

Shortly after the Denver Project started an expert advisory committee was formed. This was a requirement of the US EPA Cooperative Agreement and the Department had already used this process successfully during its pilot testing and for development of the plant conceptual design. Members of this committee included experts in many fields, some that had also served on the NRC panel, to address the scope of the Project.

Members of that committee included:

Richard Bull, PhD, Washington State University

Joseph Cotruvo, PhD, U.S. EPA, Washington, D.C.

Fred Kopfler, PhD, U.S. EPA, Stennis Space Center, Mississippi

I.H. Suffet, PhD, Drexel University

Lymon Condie, PhD, U.S. Army Dugway Proving Ground, Utah (formerly US EPA)

Joseph F. Borzelleca, PhD, Medical College of Virginia

Robert Neal, PhD, Vanderbilt University

Paul Ringhand, US EPA, Cincinnati, Ohio

Raymond Yang, PhD, National Toxicology Program (and Colorado State University)

John Doull, PhD, M.D., University of Kansas Medical Center

Carl Brunner, Project Officer, US EPA, Cincinnati, Ohio


This committee advised the Department to conduct the lifetime whole-animal chronic health effect study and omit the *in vitro* toxicity, mutagenicity, and sub-chronic studies. They pointed out that the *in vitro* studies were known to give false positive outcomes often and that this would confound the results of the lifetime study. Further, US EPA was funding a large percentage (initially about 50%) of the lifetime animal study because they wanted this ultimate evaluation to be conducted and did not want to miss the opportunity to perform this extremely expensive assessment (over $3 million). The Denver Water Department agreed with this advice since they wanted to do everything possible to determine the safety of the reuse water they were considering for a future drinking water supply. The decision was consequently made to conduct a comprehensive lifetime whole-animal health effects study of water for the first time.

## Test Substance for the Animal Health Effects Study

Two important elements of the animal health effects study needed development: the method for the preparation of the test substance to be used and the animal testing procedures. The water to be evaluated couldn't be used directly because of the insensitivity of the test animals (rats and mice) and the maximum duration of the study (two years). Concentrates of the water needed to be used to compensate for these inadequacies.

131

The advisory committee recommended using a concentrated sample based on a factor of ten for interspecies factors, a factor of ten for individual factors, and a factor of five for inherent variability in the water sample and incomplete recovery of total organic carbon. The maximum dosage, thus, was 500 times the amount found in the original water. To ensure that some useful data could be obtained if this dose was toxic a lower dose of about one third (150 times the original water amount) was recommended.

Once the concentration factor was established focus turned to developing the best method to prepare this concentrate. Ideally a concentrated water sample to be used for health effects testing would contain all the constituents in the identical proportions found in the un-concentrated water. However, there were several problems with a sample which has these characteristics. One significant problem is that the salt content of the sample would be toxic to the test animals. This significant portion of the original sample then must be separated from the concentrate. This issue was dealt with by analyzing the water for a comprehensive list of inorganic substances and comparing the results with known acceptable concentrations found in other studies and included in potable water health regulations (Table 11-1).

This approach isn't applicable for the organic substances. Many organic compounds of concern have been found in municipal wastewater. Although numerous organics are removed by conventional wastewater treatment, some compounds are not significantly reduced. A major concern for potable water reuse was that methods of analysis were still limited. Although many organic compounds have been identified in wastewater effluent it was estimated that only about 10% of the total organic carbon is amenable to identification by chemical test methods. The list of identified compounds tended to be dominated by simpler molecules that were low molecular weight. Nonvolatile organic chemicals not normally measured with existing analytical techniques were of the greatest significance and these substances were not identified by current methods. The human health impacts of organics that have not been identified were not specifically known. However, there was concern about the potential for risk from exposure to this uncharacterized fraction of the water organic content.

Several animal bioassays had, in any case, evaluated the health effects of single chemicals. Since the list of possible organic chemicals in water is nearly endless this approach is not realistic for determining risk from a complex mixture. The NRC in *Quality Criteria for Water Reuse* recognized the limited value of a single chemical testing procedure.

The National Toxicology Program (NTP), thus, conducted a sub-chronic toxicity study on a chemical mixture. A mixture of twenty-five chemicals representing compounds frequently found in contaminated groundwater was used. Nineteen of these compounds were organic substances. The NTP protocol outline noted that most previous studies had been conducted with only one or two chemicals. This study was useful to provide guidance on how toxicity studies should be conducted on complex mixtures but it did not answer the question regarding possible health effects from naturally occurring but uncharacterized organic substances.

The Health Effects Advisory Committee concluded that the test substance for the animal health effects study should expose subject animals to concentrates of the organic chemical constituents found in the water under evaluation. The Reuse Demonstration Project included a comprehensive analytical testing program which comprised complete chemical, microbiological, and physical examinations. The primary focus for animal health effects study samples became the organic compounds which may be present but which were not identified by current analytical procedures. Depending on the specific water this fraction may account for as much as 90% of the dissolved total organic content.

## Water Concentration Procedures for Animal Health Effects Study

There were no recognized methods for concentrating all the organic substances which could potentially be found in water without altering their relative concentration or composition. For that reason when preparing water concentrates for whole-animal health effects studies one must accept trade-offs which will provide an acceptable sample to achieve the Project goals.

Large-scale water sample concentration for a two-year chronic health effects study using rats and mice had never before been attempted on drinking water. The duration of the study and the amount of sample required presented many practical obstacles. The size of the isolation columns, remote operation, pH control, material compatibility, chemical artifacts, solvent recovery, sample handling, storage, and shipping requirements were examples of the considerations examined when conducting this type of program. Optimum conditions for the consistent controlled isolation and concentration of unknown organic compounds from water needed to be determined before beginning of animal feeding study.

The acquisition of a suitable sample was complicated by several other factors including the spectrum of potential compounds which could be present, the volume of water which must be concentrated to provide the necessary sample, the lack of sufficient previous work on the collection of environmental samples for this purpose, and the probable variability in the composition and concentration of organic compounds from the treatment plant over time. These and other problems required a series of compromises when evaluating a method for isolation and concentration of organic compounds from water samples to be used for animal health effects tests.

The animal health effects study expert advisors recommended the preparation of a 500-fold organic concentrate. This necessitated the construction of concentrating apparatus on a scale never before attempted. Considerations that reduced the volume of solvents and simplified the procedure were necessary to insure success by minimizing areas which might introduce contaminants or cause an interruption in the sample flow. Many methods which could be suitable for the small volume needed for chemical testing were not practical for a two-year whole-animal chronic health effects study. On the other hand since the objective was produce samples for animal health effects testing rather than analytical testing the importance of trace chemical artifacts which may interfere with chemical analysis may not be as important as long as they do not produce toxic effects.

If at all possible, the organic concentrate should be representative with respect to the chemical species and concentration ratios of all the organic substances present in the water collected over a specified period. No single concentration procedure existed that was capable of concentrating all the organic materials present in water samples. A concession position that was usually taken was that a representative sample of organic matter be collected which included the efficient extraction of potential toxicants as well as those of unknown health significance.

An issue important in potable water evaluations was the change in the type and concentration of organic compounds over time especially in a surface water source. Both the types of compounds present and their concentrations may change. Thus it was impossible to synthesize a typical mixture of organic compounds for a biological testing program. A concentrate of a composite sample representing water over a period of time was best for the animal health effects study testing program.

This procedure tends to mask the effect of high concentrations of organics that exist only for short duration. However, this situation was not a concern for long-term health effects toxicity testing programs where exposure was averaged for a lifetime  Collecting concentrate samples frequently over the two-year study, however, would capture any seasonal or other variations and result in animal exposure equivalent to a human water customer. Also, collecting concentrate samples and using them immediately would eliminate any concerns about possible changes that might occur in the test substances upon storage.

The selection of the isolation and concentration method must consider the test organism. For example, a long-term feeding study using mice required organic material from many thousands of liters of water whereas bacterial bioassays may require only 100 L. Also, the concentration method must be chosen on the basis of the chemical and physical properties of the organic constituents being tested. A quality assurance, quality control, program to limit artifacts, thus, was required. Considering these factors, several potential methods were identified for preparing concentrates for animal health effects studies. The potential concentration methods included: liquid gas methods such as static headspace, purge and trap, closed loop stripping, distillation, and evaporation, and freeze drying methods; liquid solid systems such as resins, ion exchange, and membranes; and liquid-liquid extraction methods.

After a thorough assessment of the literature and input from the Health Effects Advisory Committee it was determined that the concentration of large volumes of water for use in chronic animal health effects studies could be achieved best by only two primary methods: absorption on XAD resins (Amberlite® Rohm and Haas) or continuous liquid-liquid extraction. Dr. Suffet was the lead chemist on the committee since he had conducted research into methods of concentrating organics from water and was a contributor to the NRC report *Quality Criteria for Water Reuse.* Dr. Kopfler and Dr. Grabbe both with US EPA provided additional advice and reviewed Dr. Suffet's recommendations and conclusions.

The XAD adsorption method was used by many researchers with varied results. The method can isolate naturally occurring organic compounds from water while also achieving retention of many synthetic organic compounds. Continuous liquid-liquid extraction holds promise when concentrating synthetic organic compounds and has a potential advantage since trace artifacts, which may be a concern with an XAD resins, were reduced.

Both of these methods required pilot testing to determine the suitability for the use on the specific water used for animal health effects preparation. After extensive pilot testing liquid-liquid absorption was rejected for use in isolating and concentrating samples for the Reuse Demonstration Project. This was primarily due to the failure of this method to isolate a significant fraction of the dissolved organic carbon content of natural water. The liquid-liquid extraction method did show excellent capability of removing synthetic organic compounds that can be identified by gas chromatography.

Thus, the sample isolation method chosen after exhaustive laboratory and pilot-scale testing for the Reuse Demonstration Project animal health effects testing program was adsorption on XAD resins. The isolation system consisted of two four-inch diameter stainless steel columns each 40 inches long. The columns were in series and the first one contained XAD 8 and the second an equal mixture of XAD 2 and 4. The influent pH to the columns was maintained at 2.0 ±0.3 using hydrochloric acid. Pressures and flows were monitored and recorded. Automatic bypass was provided to divert the flow if the pH was outside the operating range.

The resins were eluted with acetone. The organic residue was collected in glass containers and the acetone was removed by rotary vacuum evaporation at a maximum temperature of 40° C. Recovered acetone was distilled and tested for purity before recycling. The water concentrate sample was diluted with laboratory-grade high purity water to obtain a 5000 to 1 concentrate that was diluted by a factor of ten at the animal testing laboratory. The acetone level in the concentrate did not exceed 1000±500 mg/L. This resulted in a maximum 100±50 mg/L acetone residual in the high dose sample (500x). Emulphor® EL-620 was added to the sample 0.25% by weight. The Emulphor® concentration in the 500 to 1 sample water was then 0.025%. The concentrate sample was sparged with dry nitrogen and stored in the dark at 4° C. Samples were shipped to the health effects laboratory overnight at 4° C in insulated boxes. Small aliquots were taken from the concentrates and tested for quality assurance. Analyses were performed to verify the pH, solids, tannin and lignan, chloride, conductivity, Emulphor®, and acetone level before shipping.

The health effects advisory committee reviewed the support data and the results from pilot-scale evaluations and affirmed the XAD resin isolation method as the best procedure for preparing organic concentrates for animal health effects testing. The method was capable of recovering a wide variety of model compounds and large percentage of the unknown dissolved organic fraction. Although many solvent elution systems were evaluated, and some exhibited certain advantages, acetone provided on balance, excellent performance along with simplicity and low toxicity. One change in the procedure that was suggested by the committee was to add back the volatile organic compounds lost in the procedure. This suggestion was adopted, but the procedure had to be performed at the animal health effects study laboratory site due to the volatility of these compounds.

## Animal Health Effects Study Procedures Overview

The Reuse Demonstration Project received the generous advice from many toxicologists and experts in chemistry, microbiology, and engineering to develop the test procedures for the animal health effects study. These were written by Dr. Joseph Borzelleca, Professor of Toxicology at Medical College of Virginia, and Dr. Lymon Condi, toxicologist with US EPA. The remainder of the health effects advisory committee and the Project advisory committee reviewed these draft protocols and provided input that was included in the final version.

The basis for the testing program was the National Toxicology Program *General Statement of Work for the Conduct of Toxicology and Carcinogenicity Studies in Laboratory Animals.* The Cooperative Agreement with US EPA required that the testing protocol include *Good Laboratory Practice* regulations from that agency, so these were incorporated in the Project procedures. The general protocol called for comparative testing of water concentrates obtained from Denver drinking water, Reuse Plant product water using reverse osmosis as one of the treatment steps, and reuse product water substituting ultrafiltration for reverse osmosis. The comprehensive whole-animal health effects study consisted of three separate but related studies: a 104-week chronic toxicity and carcinogenicity study using Fisher 344 rats: a 104-week chronic toxicity and carcinogenicity study using $B_6C_3F_1$ mice; and, a two-generation reproductive toxicity study using Sprague-Dawley rats.

Two dosage groups per water sample were administered to two species, both rats and mice, for the chronic toxic toxicity and carcinogenicity studies. The two dosages were established for this study was 500 times and 150 times the concentration in the original water samples. The sample quantity from the ultrafilter process sequence was limited due to the size of the process components so only the highest dose, 500 times, was used in only the rat species. Distilled water was used as the control. The reproductive toxicity study used only the high dose since any adverse health effect would result in a rejection of potable reuse as a water supply alternative.

The water concentrates were diluted with distilled water and administered to the test animals as drinking water. The water samples contained 0.025% Emulphor® and 100 mg/L acetone. Amber glass sipper bottles with Teflon® stoppers and stainless steel tubes were used for all animal feeding studies. The Denver drinking water samples contained precisely measured volatile organic substances added to the concentrate samples because they were lost during the sample preparation procedures (Table 12-1).

The water sample concentrates were isolated on XAD resins and organic substances were eluted using acetone. Rotary vacuum distillation was used to remove the acetone from the water residue. A five thousand fold concentrate was prepared using distilled water and shipped to the animal health effects study laboratory (located in Rockford, Maryland). The concentrates were shipped in amber glass bottles packed under nitrogen and kept at 4° C. The animal testing laboratory used distilled water to dilute the concentrates and added the volatile organic compounds for the Denver drinking water samples before administering the diluted concentrates as drinking waters for the test animals. Unused concentrates were resealed under nitrogen and stored at 4° C.

The concentrated water samples used for the animal studies were prepared continuously and simultaneously as the studies were being conducted. Preparing samples frequently over the entire two-year animal testing study resulted in variable exposure like customers drinking reuse water. This was also required to avoid any concern about changes that might occur upon storage. The sample preparation equipment, thus, was sized so that it could keep ahead of the demand but could not produce excess quantities to operate more than a few weeks without needing additional concentrate. Upon completion of the animal health effects study more than five million gallons of water had been processed to produce the sample concentrates.

**Table 12-1**
**Volatile Organic Chemicals (VOCs)**
**Added to Denver Water Concentrate Samples**

| Compound | Concentration (mg/L) 500x sample/150x sample |
|---|---|
| Chloroform | 1.5/0.45 |
| Bromodichloromethane | 0.5/0.15 |
| 1,1-Dichloropropanone | 0.2/0.06 |

*104-Week Chronic Toxicity and Carcinogenicity Studies*

The NTP protocol used for the chronic studies examined potential effects on growth, development, and carcinogenic effects. The procedure called for seventy (70) males and seventy (70) females of both Fisher 344 rats and $B_6C_3F_1$ mice and for each dosage and test sample. The animals were 6 to 11 weeks of age at the start of the study. Animals were selected for each test group by formal randomization procedures. The total number of animals at the start of this study was 560 for the reuse effluent sample from the treatment sequence including reverse osmosis, 560 animals for Denver drinking water sample, and 140 animals for the reuse effluent sample from the treatment sequence including ultrafiltration, and 280 animals in the control group, for a total of 1,540 animals.

Animals were evaluated for: clinical observations, survival rate, growth, food and water consumption, organ weights, gross necropsy, and histopathological examination of all lesions and major tissues and organs (Table 12-2), hematology, and clinical chemistry (Table 12-3).

**Table 12-2**
**Tissues for Histopathologic Evaluation**
NTP Protocol for Chronic Animal Studies

| | |
|---|---|
| Gross lesions and tissue masses (and regional lymph nodes) | Heart |
| Mandibular and mesenteric lymph nodes | Esophagus |
| Bronchial nodes | Stomach (forestomach and glandular stomach) |
| Salivary gland | Uterus |
| Femur, including marrow | Brain (three sections, including frontal cortex and basal ganglia, parietal cortex, and thalamus, and cerebellum |
| Thyroid gland | Thymus gland |
| Parathyroid glands | Larynx |
| Small intestine (duodenum, jejunum, ileum) | Trachea |
| Large intestine (cecum, colon, rectum) | Pancreas |
| Liver | Spleen |
| Gall bladder (mouse) | Kidneys |
| Prostate | Adrenal gland |
| Testes/epididymis/seminal vesicle | Urinary bladder |
| Ovaries | Pituitary gland |
| Lungs and mainstem bronchi | Spinal cord and sciatic nerve (if neurologic signs were present) |
| Nasal cavity and nasal turbinates (three sections) | Eyes (if grossly abnormal) |
| Preputial or clitoral glands | Mammary gland |
| Pharynx (if grossly abnormal) | Skin |

**Table 12-3**
**Chronic Animal Study**
**Hematology and Clinical Chemistry Evaluations**

| Hematology | Clinical Chemistry |
|---|---|
| Erythrocyte count | Sorbitol dehydrogenase (SDH) |
| Mean corpuscular volume | Alkaline Phosphatase (ALP) |
| Hemoglobin | Creatine Kinase (CK) |
| Packed cell volume | Creatinine |
| Mean corpuscular hemoglobin | Total Protein |
| Mean corpuscular hemoglobin concentration | Albumin |
| Erythrocyte morphologic assessment | Urea Nitrogen (BUN) |
| Leukocyte count | Total Bile Acids |
| Leukocyte differential | Alanine Aminotransferase (ALT) |
| Reticulocyte count | Glucose |
| Platelet count and morphologic assessment | Cholesterol |
| | Triglycerides |

*Reproductive Toxicity Study*

Reproductive performance, intra-uterine development, and growth and development of the offspring during two generations were evaluated for the reproductive toxicity study. Potential teratology effects were examined for fetal observations including variations and malformations of external, skeletal, and soft tissues.

The initial generation for the study used 50 male and 50 female Sprague-Dawley rats for each high dose water sample. Computer generated randomization procedures were used to select animals for each test group. The animals were 12 to 15 weeks of age at the start of the study.

The breeding procedures for the first and second generations followed the NTP protocol for reproductive toxicity studies. All animals were observed two times daily at least six hours apart for behavior, morbidity, and mortality. Body weight was taken weekly except during the mating. Reproductive indices and gross and necropsy and histopathology of reproductive organs of the initial and first generation parental animals examined all important tissues and major organs. Gross necropsy was performed on all parental animals dying during the treatment. Microscopic examination was made of all tissues showing gross pathological changes. Terratological examination was conducted on all appropriate samples. Fetal findings were classified as malformations or developmental variations.

## Animal Health Effects Study Results

The results from the two-year chronic toxicity and carcinogenicity study in rats and mice, and the reproductive study were first evaluated by the toxicologists at the health effects testing laboratory (Hazelton Laboratories America, Vienna, VA). These findings were subsequently reviewed and assessed by Dr. Lymon Condie, US EPA Health Effects Laboratory and Dr. Joseph Borzelleca, Department of Toxicology and Pharmacology, Medical College of Virginia. Dr. Condie prepared reports on these study findings that were then reviewed by the Project Health Effects Advisory Committee before they were submitted to the US EPA and the entire Project Advisory Committee. The findings were confirmed by these expert committees and the reported results were published in technical peer review journals (*Journal of Toxicology and Environmental Health)* and in the Project final report delivered to US EPA (*Final Report* by William C. Lauer for USEPA Cooperative Agreement No. CS-806821-01-4 April 1993).

### *Two-Year Chronic Toxicity and Carcinogenicity Rat Study*

*No toxicological or carcinogenic effects were found in Fisher 344 rats* resulting from the administration for at least 104 weeks of concentrated drinking water obtained from treated wastewater or from Denver drinking water.

The samples were concentrated up to 500 times the amount found in the original water samples. Three water samples were used in the study: unchlorinated secondary treated wastewater subjected to reuse treatment including reverse osmosis, unchlorinated secondary treated wastewater subjected to reuse treatment including ultrafiltration substituted for reverse osmosis, and Denver drinking water.

### *Study Observations*

- The study procedure required a gross necropsy examination for all test animals. No treatment related gross lesions were found. Dead animals during the study had many lesions, but these were found sporadically and in all test groups. Animal survival rates for both males and females were normal for each group.

- Animals receiving Denver tap water samples exhibited small and inconsistent but statistically significant variations in body weight, food consumption, and water consumptions. A slight taste due to the addition of naturally occurring volatile organic chemicals to this sample was likely the cause. These compounds were added to the test sample because they were lost in the concentration procedure. No volatile organic compounds were added to the reuse treatment samples since they were not detected in the original water.

- Clinical and gross pathology examinations at weeks 26, 52, and at termination did not detect any treatment related findings. A wide variety of spontaneously occurring incidental lesions were observed but these spontaneous neoplasms were observed in all groups (Table 12-4). These were the type and frequency anticipated in this age and strain of rat.

- A slightly higher incidence of "C" cell adenoma of the thyroid gland was observed in Denver drinking water males and reverse osmosis females. The variety, frequency, and severity of spontaneously occurring incidental lesions and neoplasms were all within the anticipated range, and so were not treatment related.

**Table 12-4**
**Chronic Rat Study**
**Number of Animals with Neoplasms at Sacrifice (terminal and unscheduled)**
Approximately 50 animals examined in each group and sex

| Tissue Examined | Control Group Male/Female | RO Reuse Sample Male/Female | UF Reuse Sample Male/Female | Denver Water Male/Female |
|---|---|---|---|---|
| **Pancreas** | | | | |
| Islet Cell Adenoma | 3/0 | 4/0 | 3/1 | 1/2 |
| Islet Cell Carcinoma | 1/0 | 0/0 | 2/0 | 1/1 |
| **Mammary Gland** | | | | |
| Fibroadenoma | 1/7 | 1/5 | 0/1 | 0/10 |
| **Thyroid** | | | | |
| Follicular Cell Adenoma | 5/2 | 3/2 | 3/2 | 1/0 |
| Follicular Cell Carcinoma | 0/1 | 2/0 | 3/0 | 1/0 |
| "C" Cell Adenoma | 2/4 | 3/8 | 5/2 | 9/4 |
| "C" Cell Carcinoma | 3/2 | 1/1 | 5/2 | 4/3 |
| **Pituitary** | | | | |
| Adenoma | 19/19 | 22/19 | 19/20 | 11/22 |
| Carcinoma | 0/0 | 0/0 | 0/0 | 1/0 |
| **Adrenal Medulla** | | | | |
| Benign Pheochromocytoma | 8/3 | 8/2 | 9/0 | 4/3 |
| Malignant Pheochromocytoma | 0/0 | 0/1 | 0/0 | 0/1 |
| **Hematopoietic Neoplasia** | | | | |
| Leukemia, Mononuclear | 27/16 | 22/18 | 21/11 | 21/11 |
| **Testis** | | | | |
| Benign Interstitial Cell Tumor | 46 | 38 | 44 | 47 |
| Malignant Mesothelioma | 1 | 2 | 4 | 1 |
| **Uterus** | | | | |
| Endometrial Stromal Polyp | /7 | /7 | /6 | /4 |
| Endometrial Stromal Sarcoma | /1 | /1 | /0 | /0 |

### *Two-Year Chronic Toxicity and Carcinogenicity Mouse Study*

*No toxicological or carcinogenic effects were found in $B_6C_3F_1$ mice* resulting from the administration for at least 104 weeks of concentrated drinking water obtained from treated wastewater or from Denver drinking water. The samples were concentrated up to 500 times the amount found in the original water samples. Two water samples were used in the study: unchlorinated secondary treated wastewater subjected to reuse treatment including reverse osmosis and Denver drinking water.

### *Study Observations*

- The study procedure required a gross necropsy examination for all test animals. No treatment related gross lesions were found. Dead animals during the study had many lesions, but these were found sporadically and in all test groups. Animal survival rates for both males and females were normal for each group.

- Animals receiving Denver tap water samples exhibited small and inconsistent but statistically significant variations in body weight, food consumption, and water consumptions. A slight taste due to the addition of naturally occurring volatile organic chemicals to this sample was likely the cause. These compounds were added to the test sample because they were lost in the concentration procedure. No volatile organic compounds were added to the reuse treatment samples since they were not detected in the original water.

- Clinical and gross pathology examinations at weeks 26, 52, and at termination did not detect any treatment related findings. There was an apparent incidence of slightly increased renal tubular regeneration in males receiving the water concentrates for at least 26 weeks. Examination after week 65 and at termination did not confirm the apparent increase.

- Neoplasms (Table 12-5) were observed in the liver, lung, and pituitary mostly after the week 66. Aging mice commonly exhibit these neoplasms and there was no treatment related relationship. All the other observed microscopic changes were considered consistent in type and severity with common spontaneous processes for this species.

**Table 12-5**
**Chronic Mouse Study**
**Number of Common Neoplasms (70 animals per group and sex)**

| Organ | Control Group Male/Female | RO Sample Male/Female | Denver Water Male/Female |
|---|---|---|---|
| Liver | 28/2 | 21/5 | 24/7 |
| Lung | 10/4 | 12/2 | 11/2 |
| Pituitary | 0/10 | 0/12 | 0/9 |
| Hematopoietic System | 9/20 | 10/25 | 7/21 |

*Reproductive Toxicity Study*

*No demonstrated treatment related effects* were found related to reproductive performance, growth, mating capacity, survival of offspring, or fetal development in the multi-generational reproductive study. Three test articles were used in this study: reuse treated water using reverse osmosis, reuse treated water using ultrafiltration substituted for reverse osmosis, and Denver drinking water.

**Table 12-6**
**Reproductive Toxicity Study**
**Teratogenicity**

| Fetal Incidence Type | Control Group #/Examined | RO Sample #/Examined | UF Sample #/Examined | Denver Water #/Examined |
|---|---|---|---|---|
| Skeletal Variations | 92/127 | 79/124 | 50/79 | 82/108 |
| Skeletal Malformations | 0/127 | 1/124 | 0/79 | 0/108 |
| Soft Tissue Variations | 7/58 | 12/63 | 4/42 | 20/58 |
| Soft Tissue Malformations | 0/58 | 0/63 | 0/42 | 0/58 |
| External Variations | 0/185 | 0/187 | 0/121 | 0/166 |
| External Malformations | 0/185 | 1/187 | 0/121 | 0/166 |

*Study Observations*

- All animals in the initial generation group survived except for one female in the Denver water dose group whose death was due to a difficult delivery. This generation showed no difference in body weight gain for any treatment group.

- The daily water intake was consistently lower for the Denver drinking water dose group. Again this was probably due to the volatile organic compounds added to the Denver drinking water dosing solutions. No volatile organic compounds were added to the reuse product water solutions.

- First-generation survival was good throughout lactation and this generation's body weights were similar in all groups. No adverse effects on pup survival or growth were noted in either of the second generation pup groups.

- Only one malformed fetus was found (Table 12-6). Skeletal and visceral variations in development did not occur in a pattern that would indicate an experimental effect.

- At necropsy no clinical signs or gross tissue alterations were noted.

- Histopathological examinations in parental animals of either generation found nothing treatment related.

## Conclusions

The Denver Direct Potable Water Reuse Demonstration Project conducted, for the first time on drinking water, comprehensive lifetime animal health effect studies to evaluate the effects of consuming water reclaimed from wastewater for possible use as drinking water. These unprecedented studies included two-year chronic toxicity and carcinogenicity assessments on both rats and mice and reproductive toxicity studies on rats. Concentrated water samples (500 fold maximum concentration) obtained from reuse treated water using either reverse osmosis or ultrafiltration as part of a treatment sequence and Denver tap water were used as drinking water for the test animals.

*No treatment related health effects were found in any of the samples.* This result combined with those from the comprehensive water quality testing program established the high quality of reuse treated wastewater that not only meets all drinking water regulations but is comparable or superior to Denver's high-quality tap water.

# 13

# Public Information Program

The public information program was a unique and important aspect of the Reuse Demonstration Project. The outcome of the plant reliability and water quality testing portions of the Project were critical to gaining acceptance of direct potable reuse by the public. However, these results would not be complete until the Demonstration Project was concluded. The public information program, as a result, focused on potable water reuse education of potential customers and drinking water regulators.

Several public opinion surveys were conducted early in the Project which revealed essentially the same consumer attitude toward using reclaimed wastewater as a drinking water source. The survey results showed that the better informed people were about potable reuse the more likely they were to accept it. An extensive study funded by the Office of Water Research US Department of Interior concluded that on-site tours of the demonstration facility where the most effective method of educating and informing the public. Also, this study listed a variety of informational strategies which would increase public awareness. The Project public information program, implemented from 1979 to 1991, generally followed these suggestions.

The public information program had two interrelated but distinct objectives. The first goal was to increase Denver area resident's awareness of the direct potable reuse. The second goal was to inform regulatory agency decision-makers of the Project progress and technical results.

The first objective would be satisfied by contacting at least 50,000 Denver area residents with information regarding this potential new water supply. This number could not be accommodated in person at the Reuse Demonstration Plant. Other means were then utilized to provide information to a larger audience.

Regulatory agency acceptance was achieved by recognition of the credibility of the Project results and a realization of the need and the public demand to use this resource. It was not realistic to expect regulatory agencies to certify this technology at the conclusion of the Project. However, one element which would generate regulatory agency acceptance was the technical merits of the scientific evaluations. Thus, every opportunity was taken to share results, include regulatory agency input and review, and to conduct the test program with US EPA participation and the Colorado State Health Department surveillance thus ensuring agency awareness.

The information program utilized a multimedia approach to reach the widest possible audience. Personal contact on escorted plant tours was used whenever possible, since this was determined to have the greatest impact on attitudes. A professionally produced video was used to reach those who could not visit the site in person. Printed material was distributed to even more potential customers. Technical reports and personal appearances at scientific conferences reached regulators as well as industry leaders. The combined effect of these efforts resulted in the education about potable water reuse of a broad group of stakeholders.

A great deal of effort went into designing the plant tour since this was identified as having a great influence over public and regulator attitudes. The entire demonstration plant was designed with this in mind. The exterior was landscaped with Xeriscape® low-water use plantings in keeping with the water conservation theme of the Project. The treatment processes, except for lime clarification, were housed indoors. The plant tour route was handicapped accessible and generally followed the flow of the water through the plant. Wide aisles were provided to accommodate large groups and all the tanks and interconnecting piping were color-coded for easy identification. Non-technical informational displays were installed at each major treatment step to facilitate the translation of information of a highly technical nature to non-technical language. The administrative and reception areas were attractively furnished.

A tour gathering conference room was included with audiovisual facilities and seating for eighty. This area also served as a meeting room and training facility for plant personnel. A narrated slide presentation was used as an orientation device at the beginning of the tour. This ensured continuity of information and added a professional touch to the visit. Color computer terminals were provided to demonstrate real time plant operational data and allow hands on interaction by visitors with the intricate treatment system. A full-color brochure describing the plant and the research programs was provided to each visitor to share with family members and friends. The brochure was carefully written so as not to become dated and this continued to be used even after the plant was closed.

Videotaped presentations were recognized as a powerful tool for communication about the Project. Several video programs were developed to give a plant tour experience for those who could not visit in person. These programs illustrated segments of the Project, laboratory studies, and water analysis procedures that were difficult to view in person. The most important video was the professionally produced program; *Pure Water...Again*. This twenty-six minute video was produced as a documentary about the Reuse Demonstration Project and was suitable for airing on television. Schools became the largest user of this video since travel to the plant was not always possible. Also, this program was produced so as not to become dated so that it could be used for many years to explain the work conducted at the Reuse Demonstration Plant. Based on market share estimates of the program broadcasts on public television channels, more than 50,000 Denver area residents viewed this program. In addition, the community access channel aired the documentary more than twenty times. Viewership was estimated at over 10,000.

A variety of printed information was utilized to reach a diverse audience. The Project brochure, *Welcome to Tomorrow,* was provided to many interested parties besides plant visitors. The quarterly Successive Use Newsletter was developed to keep local as well as national and foreign interested parties informed of the Project progress and inform them of important milestones. The distribution list expanded from 400 to 2000 by Project end. Twenty-one issues were published over the term of the Project. Inserts in water bills were distributed on several occasions to tell Denver Water Department's one million customers about the Project.

Personal appearances at technical conferences (30), local groups, and schools were very important to establish credibility and create opportunities for information exchange. Extensive professional slide or video visuals were developed to augment the spoken narration. A great deal of interest and complementary feedback was obtained at these presentations.

Facility tours were offered shortly after construction began with the first documented visitors arriving in 1981. Nearly 7,500 people were hosted on tours conducted by Project personnel. Early in the Project most interested parties were technical having heard of the Project through the scientific literature. International visitors from 47 countries attended tours at the plant.

A technical presentation was made by video at a conference held in Kyoto Japan in 1989. This program received much recognition among professionals in the field. Also, video was used in conjunction with national conferences held in Denver on several occasions both at the conference site and in the plant to augment normal tour materials. These programs transmitted a powerful message about the extent of the research and the thoroughness of the reuse demonstration testing program.

The Project final report (25 volumes) and nine major interim reports totaling more than 15,000 pages were published and distributed to US EPA, Water Department staff, and Project advisors. These reports documented the results and provided data used to support decisions and modifications. Also, the treatment plant operations and maintenance manuals (seven volumes) and the Project assurance quality assurance manuals (eleven volumes) supported the scientific studies. In addition to these progress reports and operational support manuals more than 100 technical articles appeared in national and international journals. Federal, state, and local regulatory agencies were kept informed of the Project results through these publications as well as their memberships on Project advisory committees. The US EPA, was involved directly in the Project by providing $7 million in funding. Thus, US EPA was provided all published reports for review and approval.

News articles about the Project were carried locally and in publications from Maine to California. Various magazines mentioned the Project including National Geographic, Time, Science, Public Works and the Journal of the Freshwater Foundation. Television coverage by local CBS affiliate channel 7, independent channel 2, NBC affiliate channel 4, and public broadcasting channels 6 and 12, was joined by programs originating in Orlando, Florida, Phoenix, Arizona and Los Angeles, California where the Denver Reuse Demonstration Project was highlighted. This media coverage made the Project one of the most recognized water research studies ever undertaken.

The treatment facility and Project personnel received several awards recognizing contributions to the advancement of water treatment technology. These acknowledgments were verification of the technical community's appreciation for the accomplishments of the Potable Water Demonstration Project. Most notable of these were Engineering Excellence awards from the Consulting Engineers Council, Men Who Made Marks from Engineering News Record magazine, the outstanding engineering achievement award from the Professional Engineering Council, the certificate award for landscaping from Commerce City community pride project, and the award of merit for excellent safety record every year of operation from the Rocky Mountain Water Pollution Control Federation.

## Conclusions

The Direct Potable Water Reuse Demonstration Project public information program utilized a multimedia approach to achieve its two main objectives. Many more than the 50,000 Denver area residents targeted to receive educational information were contacted during the Project. The awareness of local populace about the possibility of potable water reuse use was definitely increased during the Demonstration Project and the protocol was established to continue this educational program to increase public awareness.

Regulatory agencies were involved in every aspect of the Project. The US EPA was a funding contributor and was a partner in all decisions, received all reports, and reviewed all results. The Colorado State Health Department as well as local health authorities were part of the Project advisory committee and as such received all technical communications provided to the US EPA. Members of these agencies as well as the engineering and scientific community at large were informed about the Project results through hundreds of technical and news reports. This unprecedented communication effort established Project credibility and assured regulatory agency awareness.

# 14

# Project Conclusions

The Direct Potable Water Reuse Demonstration Project was initiated in 1979 with the signing for a Cooperative Agreement between the Denver Water Department and the US EPA. This followed more than ten years of preliminary pilot-scale studies and planning for a demonstration-scale treatment facility. The Project was developed to determine the economic and technical feasibility of reliably producing potable quality water from unchlorinated secondary treated wastewater. A 1 mgd demonstration treatment plant was designed, constructed, and operated providing a unique testing facility to conduct comprehensive evaluations. Water quality was of primary concern and thus, the Project included comprehensive water quality analyses and a lifetime whole-animal health effects study. The Project also included a public information program and regulatory agency involvement to raise awareness about the possibility of direct potable reuse to meet Denver's future water needs. The primary findings of the ten-year Project are:

- The reuse treatment process sequence reliably produced water satisfying all current and proposed US EPA and all international drinking water standards.

### Treatment Sequence

1. High pH lime clarification
2. Recarbonation
3. Filtration
4. Ultraviolet Irradiation
5. Activated carbon adsorption
6. Reverse osmosis or Ultrafiltration
7. Air stripping
8. Ozonation
9. Chloramination

- The water quality produced by the reuse treatment plant was superior or equal to Denver's current drinking water. A 50/50 blend of water treated either by reverse osmosis or ultrafiltration compared favorably with Denver drinking water.

- More than six years of operation and special studies designed to challenge the treatment plant have shown the treatment system capable of removing all contaminants of concern

153

and to provide a level of protection from pollutants well beyond conventional water treatment facilities.

- No compound, substance, or organism was found in the reuse product water in an amount of any concern. The reuse product water quality was found to be superior to most drinking water supplies.

- A whole-animal (two species) lifetime chronic toxicity and carcinogenicity study was conducted for the first time on drinking water. This study found no adverse health effects in either the reuse water produced through reverse osmosis or ultrafiltration treatment systems, or in the Denver drinking water used as a comparison. The complementary whole-animal multi-generational reproductive toxicity study reached the same conclusion.

- The public information program was successful in raising awareness of direct potable reuse. Attitude survey results found that the majority of the Denver residents would accept potable reuse if the need was demonstrated and the safety was assured.

- Regulatory agencies including US EPA, State, and local health agencies were included on the Project expert advisory panels where they provided expert advice, participated in the assessment of the Project, and approved the final reports that presented the results.

- The cost of direct potable reuse treatment was found to be comparable to developing future conventional water supply projects propose for Denver.

~ ~

The Direct Potable Water Reuse Demonstration Project was successfully completed. The water quality produced from treated wastewater was shown to satisfy every measure of safety including the results from an unprecedented whole-animal lifetime health effects study. As a result, direct potable reuse was available as one possible alternative for consideration, along with conventional water supply development projects, to meet future water needs. The Project's success formed a basis for other water systems considering potable water reuse both in the U.S. and around the world.

# References

Lauer, W.C., and Rogers, Stephen E., (1994) The Demonstration of Direct Potable Water Reuse: Denver's Pioneer Project. Proceedings 1994 Water Reuse Symposium, Dallas, TX.

Lauer, W.C., and Work, S.W. (1984) Denver's Potable Water Reuse Demonstration Project. National Conference on Environmental Engineering, Boulder, Colorado.

Work, S.W. and Hobbs, N.M. (1976) Management Goals and Successive Use. *JAWWA*, 68:2:86.

USEPA (1980) Report on Workshop Proceedings: Protocol Development - Criteria and Standards for Potable Reuse and Feasible Alternatives. Washington, D.C.

National Research Council. (1982) Quality Criteria for Water Reuse. National Academy Press, Washington, D.C.

Lauer, W.C., Rogers, S.E., LaChance, A.M., and Shuck, D.L. (1989) Denver's Potable Water Reuse Demonstration Project Preliminary Process Evaluations -Process Selection for Comprehensive Health Effects. Denver Water Department.

Lauer, W.C., Johns, F.J., Wolfe, G.W., Myers, B.A., Condie, L.W., and Borzelleca, J.F. (1990). Comprehensive Health Effects Testing Program for Denver's Potable Water Reuse Demonstratrion Project. Jour. Toxicology and Environ. Health 105:EE4:675.

Lauer, W.C., LaChance, A.M., Shuck, D.L., Suffet, I.H., Croy, R. (1989). Water Sample Preparation for Comprehensive Health Effects Testing Denver Water Department.

National Toxicology Program, General Statement of Work for the Conduct of Toxicity and Carcinogenicity Studies on Laboratory Animals. April 1987 Revision, NIEHS, NTP, Research Triangle Park, N.C. (1987).

U.S. Environmental Protection Agency, Good Laboratory Practices. Toxic Substances Control Act (TSCA), 40 CFR Part 792.

Lauer, W.C. (1993) Denver's Direct Potable Water Reuse Demonstration Project Final Report. (25 volumes), Denver Water Department. (executive summary available from Google books and entire report in Denver Water archives).

# Photo Gallery

The author provided most of these photos from his own collection. Several were used in the Project orientation slide presentation described in the Public Information chapter and some appeared in the informational brochure used to accompany tours. The construction photos came from the Denver Water archives and are presented here with permission from Denver Water. A few photos came from former Project staff members. All photos are shown with permission and copyright release from Denver Water, individual Project staff members, and William C. Lauer.

**Process Tanks During Construction**

**Filter Tank Installation During Construction**

**Project Advisory Committee Visiting Construction Site**

**Reuse Demonstration Plant Exterior**

**Display Fountain**

**Reception Desk and Tour Gathering Area**

**Filter Tanks**

**Lime Clarifier**

**Exterior Tanks**

**Filter and Clino (Ion Exchange) Tanks Outside Main Process Building**

**Main Process Building Interior**

**ARRP Process Tanks**

**Control Room with Annunciator Panel Above**

**Control Room Showing Entire Panel**

**Control Room Showing Operator Computer Screen**

**Chlorine Dioxide Process Room Exterior**

**Ozone Process Room Exterior**

**Reverse Osmosis Process with Informational Display**

**Air Stripping Tower with Control Room in Background**

**Tour Group Near Carbon Process**

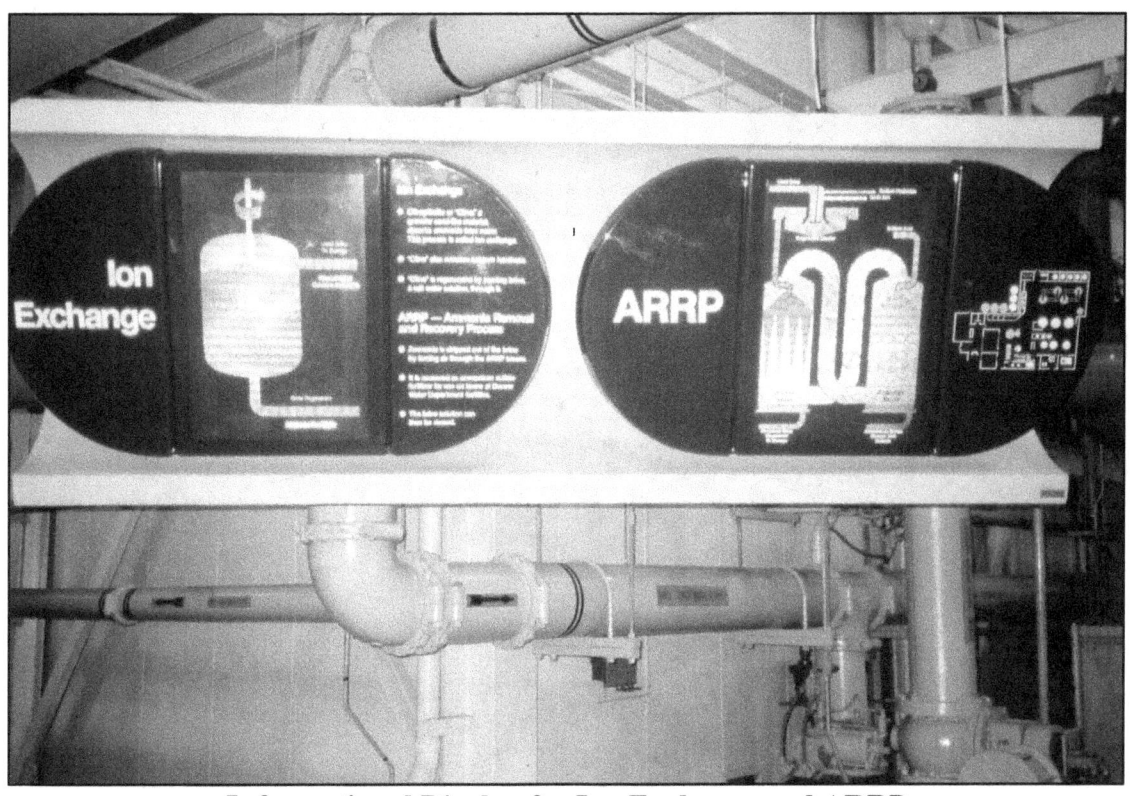

**Informational Display for Ion Exchange and ARRP**

**Plant Laboratory**

**Virology Laboratory (remote site)**

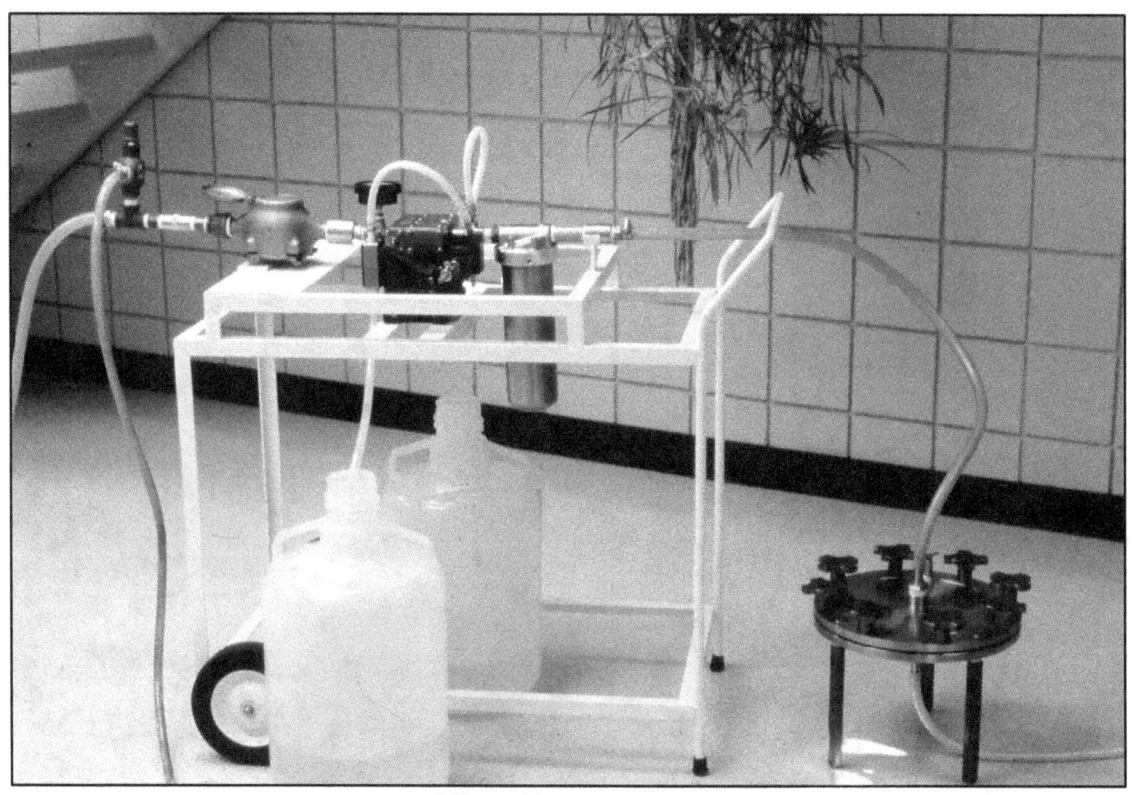

**Virus Concentrator**

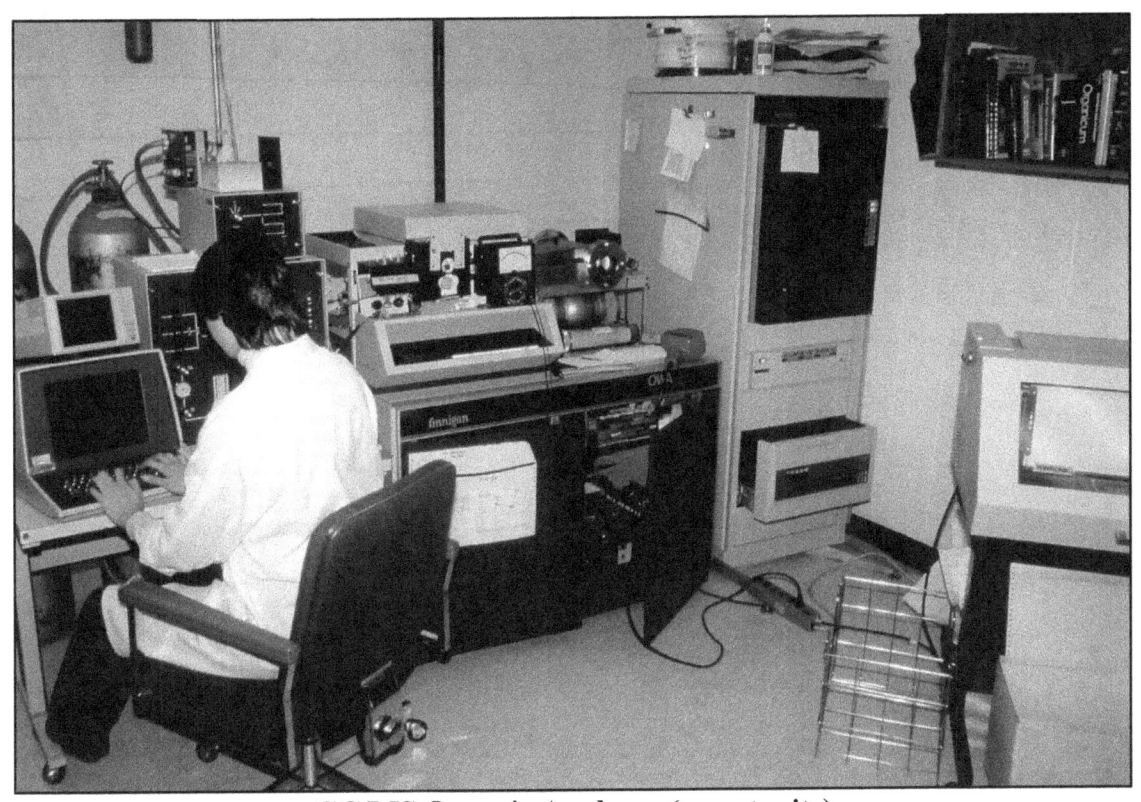

**GC/MS Organic Analyzer (remote site)**

**Animal Health Effects Isolation Columns**

**Animal Health Effects Isolation Columns at Ultrafiltration Pilot-Scale Site**

**Rotary Vacuum Distillation Equipment for Animal
Health Effects Study Samples**

**Extraction Rack for Animal Health Effects Sample Preparation**

**Main Process Area Interior- Carbon Tanks**

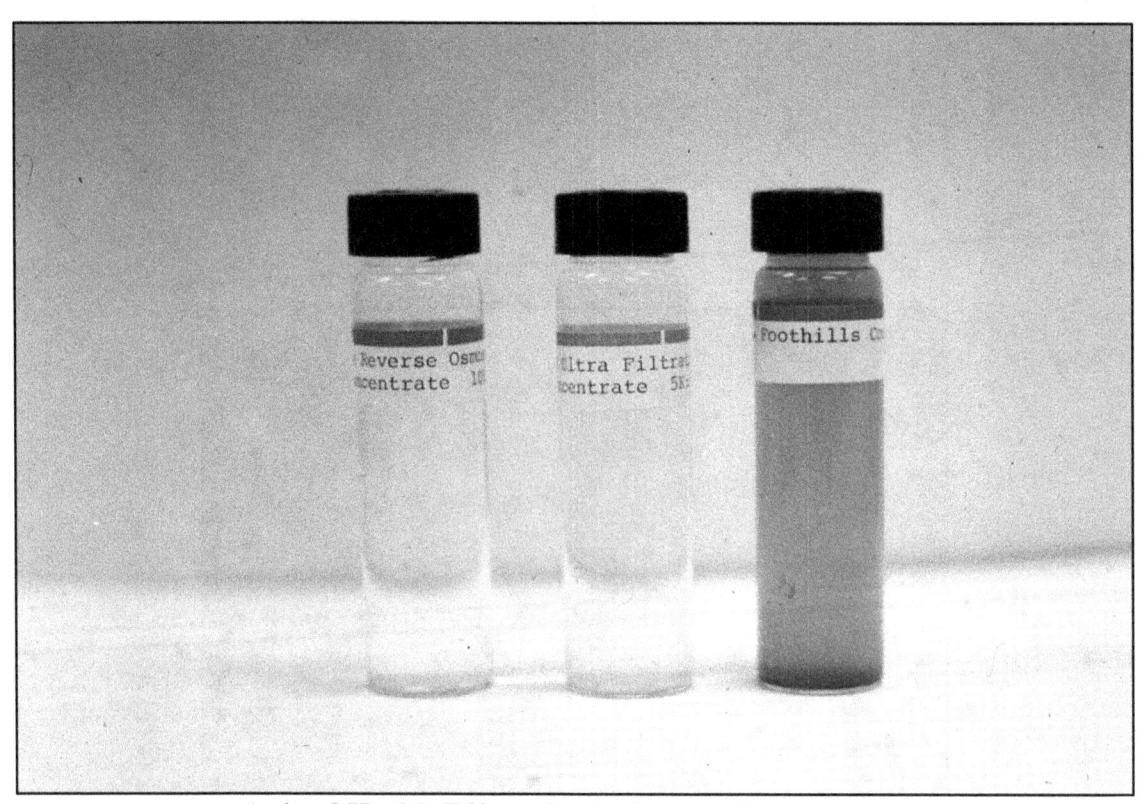

**Animal Health Effects Study Sample Concentrates**
**(Reverse Osmosis left, Ultrafiltration middle, Tap Water right)**

**Project Staff In Front of Control Room at Conclusion**

**Final Report**

# About the Author

**William C. Lauer** is an internationally recognized authority on treatment methods, drinking water quality, distribution system operations, drinking water health effects, and water reuse. Authoring and editing more than 25 technical books and 70 articles on these subjects have established Mr. Lauer as a leader in the drinking water profession. He thus served as an expert advisor to the Environmental Protection Agency, the National Aeronautics and Space Administration (NASA) about drinking water reuse in the International Space Station; the National Academy of Science, and the Government of Singapore NeWater potable reuse program.

Lauer was the Project Manager for the ten-year, $30 million Denver Direct Potable Water Reuse Demonstration Project. In this position he directed the Project to evaluate the feasibility of converting wastewater to drinking water to augment Denver's water supply. The main testing facility for this study processed one million gallons per day and was described as "the world's most complex water treatment plant." To establish water safety, a comprehensive testing program analyzed every known contaminant. As an ultimate test of the water safety the Project included a lifetime whole-animal health effects study, normally conducted to assess substance carcinogenicity or before approving new pharmaceuticals for use, to evaluate the possible effects from drinking recycled water long-term. The results of these evaluations concluded that reusing wastewater as a drinking water source was both economical and safe when compared to Denver's high quality drinking water and all recognized drinking water health standards.